Lars Mytting

The Sixteen Trees of the Somme

Translated from the Norwegian by
Paul Russell Garrett

MACLEHOSE PRESS
QUERCUS · LONDON

First published in the Norwegian language as *Svøm med dem som drukner* by Gyldendal Norsk Forlag A.S. in 2014

First published in Great Britain in 2017 by MacLehose Press

This paperback edition published in 2018 by
MacLehose Press
An imprint of Quercus Publishing Ltd
Carmelite House
50 Victoria Embankment
London EC4Y 0DZ

An Hachette UK company

This translation has been published with
the financial support of NORLA.

A CIP catalogue record for this book is
available from the British Library.

ISBN (MMP) 978 0 85705 606 1
ISBN (Ebook) 978 0 85705 605 4

1 3 5 7 9 10 8 6 4 2

Designed and typeset in 11½/14½pt Adobe Caslon by Patty Rennie
Printed and bound in Great Britain by Clays Ltd, Elcograf S.p.A.

Kirklees
COUNCIL

Library and Information Centres

Red doles Lane

Huddersfield, West Yorkshire

HD2 1YF

This book should be returned on or before the latest date stamped below. Fines are charged if the item is late.

| 4.12.19 | | |
| 28 FEB 2020 | | |

You may renew this loan for a further period by phone, personal visit or at www.kirklees.gov.uk/libraries, provided that the book is not required by another reader.

NO MORE THAN THREE RENEWALS ARE PERMITTED

The Sixteen Trees of the Somme

Also by Lars Mytting in English translation

NON-FICTION

Norwegian Wood (2015)
The Norwegian Wood Activity Book (2016)

And take me disappearing through the smoke rings of my mind
Down the foggy ruins of time, far past the frozen leaves
The haunted, frightened trees, out to the windy beach
Far from the twisted reach of crazy sorrow

BOB DYLAN, "Mr Tambourine Man"

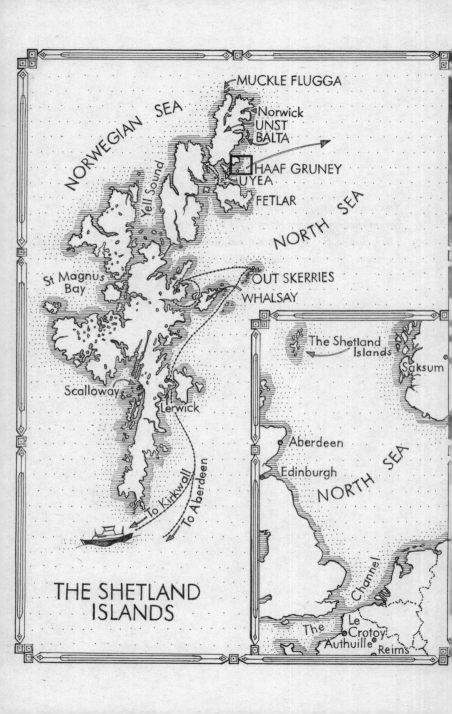

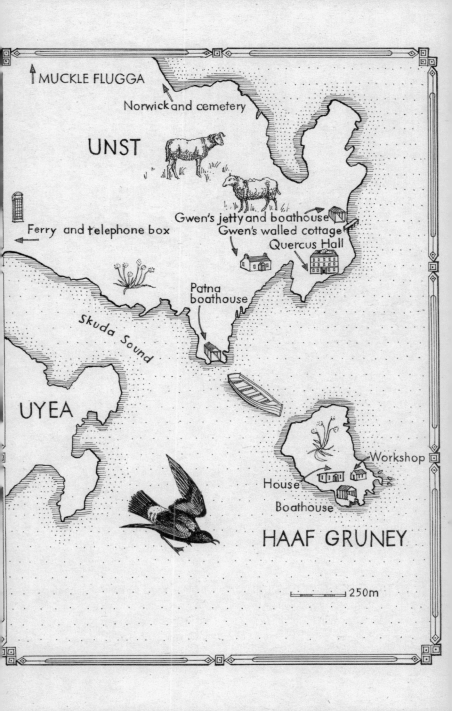

↑ MUCKLE FLUGGA

Norwick and cemetery

UNST

Gwen's jetty and boathouse
Gwen's walled cottage
Quercus Hall

Ferry and telephone box

Patna
boathouse

Skuda Sound

UYEA

Workshop

House

Boathouse

HAAF GRUNEY

250m

I

Like Ashes in the Wind

FOR ME MY MOTHER WAS A SCENT. SHE WAS A WARMTH. A leg I clung to. A breath of something blue; a dress I remember her wearing. She fired me into the world with a bowstring, I told myself, and when I shaped my memories of her, I did not know if they were true, I simply created her as I thought a son should remember his mother.

Mamma was the one I thought of when I tested the loss inside me. Seldom Pappa. Sometimes I asked myself if he would have been like all the other fathers in the district. Men in Home Guard uniforms; in football trainers at old boys' practice; getting up early at the weekends to volunteer at Saksum's local association of hunters and anglers. But I let him fade away without regret. I accepted it, for many years at least, as proof that my grandfather, Bestefar, had tried his best to do everything Pappa would have done, and that he had in fact succeeded.

Bestefar used the broken tip of a Russian bayonet as a knife. It had a flame-birch handle, and that was the only real carpentering he had ever done. The top edge of the blade was dull, and he used that to scrape off rust and to bend steel wire. He kept the other side sharp enough to slice open heavy sacks of agricultural lime. A quick thrust and the white granules would trickle out of their own accord, ready for me to spread across the fields.

The sharp and the dull edges converged into a dagger-like point, and with that he would dispatch the fish we caught on Lake Saksum. He would remove the hook as the powerful trout flapped about, furious to be drowning in oxygen. Place them over the gunwale, force the tip of the blade through their skulls and boast about how broad they were. It was always then that I would raise the oars to watch the thick blood trickle down his steel blade, while thin drops of water ran down my oars.

But the drops flowed into each other. The trout bled out and became our fish from our lake.

On my first day of school I found my way to my desk and sat down. On it was a piece of card folded in half, with EDVARD HIRIFJELL in unfamiliar writing on both front and back, as though not only the teacher but I, too, had to be reminded of who I was.

I kept turning to check for Bestefar, even though I knew he was there. The other children already knew each other, so I just stared ahead at the map of Europe and the wide chalkboard, blank and green like the ocean. I turned once more, concerned that Bestefar looked twice as old as the other parents. He stood off to one side in his Icelander, and he was old like Fridtjof Nansen on the ten-kroner note. He had the same moustache and eyebrows, but the years did not weigh heavy on him, it was as though they multiplied one another and made his face look full of vigour. Because Bestefar could never get old. He told me that. That I kept him young and that he made himself young for me.

My mother's and father's faces never grew older. They lived in a photograph on the chest of drawers next to the telephone. Pappa wearing flares and a striped waistcoat, leaning against the Mercedes. Mamma crouching to pet Pelle, our farm dog. The dog looks as though it is blocking her path, as if it does not want us to leave.

Maybe animals sense these things.

I am in the back seat, waving, so the photograph must have been taken the day we left.

I try to convince myself that I remember the drive to France, the smell of the hot imitation-leather seats, trees flashing past the side windows. For a long time I thought I could remember Mamma's distinct scent on that day, as well as their voices above the racing wind.

We still have the negative of that picture. Bestefar did not send the film for processing straight away. At first I thought it was to save money, because after this last photograph of Mamma and Pappa, Christmas Eve, midsummer net fishing and the potato harvest were still to come.

It was not obvious at the time, but I think he waited because you never know how a picture is going to turn out, not until it comes back from the lab. You have an idea, an expectation of how the subjects will settle, and within the emulsion, Mamma and Pappa would live a little longer, until the developing bath fixed them for ever.

I believed Bestefar when, as my tantrums came to an end, he repeated that he would tell me everything when I was "big enough". But maybe he failed to notice how much I had grown. So I discovered the truth too early, and by then it was too late.

It was soon after the beginning of Year Three. I cycled down to the Lindstads' farm. The door was open, I called out a greeting. The house was empty, they were probably out in the barn, so I went into the living room. A stereo and a record player stood on the dark-stained bookcase, collecting dust. Norwegian Automobile Federation road atlases, condensed novels from Reader's Digest and a row of burgundy yearbooks, with *Det Hendte* in gold letters on their spines. Each contained a summary of the most significant events of that year.

It was no coincidence that I selected the one marked 1971, it was as if the yearbook wanted me to look inside; it fell open on the month of September. The pages were shiny with fingerprints. The edges of the pages were worn and there were threads of tobacco in the gutter.

Mamma and Pappa, one photograph of each of them. Two simple profiles with "(Reuters)" printed under their names. I wondered who Reuters was, and thought I ought to know since it was about my parents.

It said that a Norwegian–French couple, "both domiciled in Gudbrandsdalen", had died on September 23 while on holiday near Authuille by the Somme in northern France. They had been visiting a fenced-off First World War battlefield and had been found dead in a river. The autopsy revealed that they had been exposed to gas from an unexploded shell, and had then lost their footing and stumbled into the water.

The yearbook went on to state that there still were several million tonnes of explosives along the old front lines, and that many areas were judged impossible to clear. At least a hundred people, tourists and farmers, had been killed in recent decades by stepping on unexploded shells.

I knew this already from Bestefar's economical explanation. The part that he had omitted came next:

"From items discovered in their car, the police established that the couple had a child with them, a three-year-old boy." But he was nowhere to be found and a search party was organised. Dogs helped to scour the former battlefield with no success, while divers dragged the river and helicopters were deployed to widen the search.

Then I read the sentences that extinguished the child in me. It was like putting newspaper in the fireplace; the writing was still legible despite the paper catching fire, but with the lightest contact it would crumble to ashes.

"Four days later the child was found at a doctor's surgery 120 kilometres away, in the seaside town of Le Crotoy. A police investigation yielded no answers. It was assumed that the boy had been abducted. With the exception of minor injuries, he was unharmed."

Then the article returned to the truth I knew, that I had been adopted by my grandparents in Norway. I stared at the pages. Flipped forward to see if anything came after, flipped back to see if I had missed anything that came before. I picked out the bits of tobacco from the gutter. People had talked about me, taken out *Det Hendte 1971* when the neighbours were over for coffee, recalled the time someone from the Hirifjell clan had made it into the papers.

My anger had a long way to go. Bestefar said he had told me everything he knew, so I carried my questions into the flame-birch woods opposite the farm. Why had Mamma and Pappa taken me to a place filled with unexploded shells? What were they even doing there?

The answers were gone, Mamma and Pappa were gone like ashes in the wind, and I grew up at Hirifjell.

Hirifjell lies on the far side of Saksum. The larger estates are on the other side of the river, where the snow melts early and the sunshine caresses the log walls and the squirearchs who live within them. It is never called the true side, occasionally the sunny side, but most often nothing at all, because only the far side needs a name for its location. Between us flows the Laugen River. The mist rising from the river is the border we must cross when we go to secondary school or shop in the village.

The far side lies in shade for most of the day. There are jokes about the people who live here, that they fire their Krags at the travelling fishmonger's van and tie together the shoelaces of drunkards who fall asleep under the haycocks. But the point is that even if you come from an estate in Saksum, you cannot adopt Parisian habits, let alone those from Hamar. *Norge Rundt* has never broadcast a bulletin from Saksum. You find the same things here that you find in other villages. The agricultural cooperative, the draper's, the post office and the general store. A road high above the village where the ambulance always gets stuck. Peeling wooden houses inhabited by people who fiddle their taxes.

Only the telephone engineers and the local agricultural officers knew that the sun actually shone all day where we lived. Because Hirifjell is situated just at the point where the hillside slopes back down. The sunny side on the inside of the far side. A garden in the woods, closed off by a gate, wherein we kept ourselves to ourselves.

Bestefar liked to sit up all night. I would lie on the sofa while he smoked cigarillos and rearranged his books and vinyls. Bach's cantatas, boxes of Beethoven and Mahler symphonies conducted by Furtwängler or Klemperer. Shelves full of books, some tattered, some new. There were so many slips of paper sticking out of *Andrees' World Atlas* and *Meyer's Konversations-Lexikon* it looked as though new pages were growing from them.

This is where I fell asleep at night, in a haze of music occasionally interrupted by the crackle of his lighter, and, half asleep, I would hear him put down *Der Spiegel* and then feel his arms around me. When I peeped through half-open eyelids, the walls and ceiling were spinning, as if I were the needle of a compass, before he laid me down, straightened my arms and legs and pulled the duvet over me. Every morning his face was there; with the light in the hall shining on his stubble and his curved moustache, yellowed from smoking, he stood with a smile revealing that he had been watching me before I woke.

The only unreasonable thing he did was to refuse to let me collect the post. When it was late it messed up his routine, and every morning at eleven he would watch for the red car up by the county road. Later he switched to having a P.O. box down in the village, saying that some strangers had forced open the lock on the postbox.

Enclosing postage stamps, I sent off for catalogues for do-it-yourself speaker kits, Schou's hunting weapons, the A.B.U. catalogue, photography equipment, fly-tying materials – I learned more from them than from any of my textbooks. My only contact with the outside world came through Bestefar, heavy envelopes on a warm car seat from his trips to the village. It was like that for an eternity until the year he returned from the annual meeting of the Association of Sheep and Goat Breeders and, out of nowhere, said that we should switch back to the postbox, it was such an effort going down to Saksum to collect every-thing.

Even before then I remember us sawing off the stock of Pappa's shotgun and going duck hunting. It was a 16-calibre Sauer & Sohn side-by-side, given to him for his confirmation, although in all likelihood he had never used it. As I grew, we glued slivers of the severed stock back on, and when my own confirmation came the orangey-brown walnut stock was marked with thin rings, a measure of my childhood with Bestefar.

But I knew that if a spruce grew too quickly, it formed wide growth rings, and when the tree was big enough for the wind to take hold, it would snap.

All my life I had heard a whistling coming from the flame-birch woods, and one night in 1991 it grew into a wind that made me falter. Because there was something about Mamma and Pappa's story that was still stirring, quietly, like a viper in the grass.

II

Summer Solstice

1

THAT NIGHT DEATH RETURNED TO HIRIFJELL. IT WAS
obvious who was going to be collected, because there were few
to choose from. I was twenty-three years old, and when later I
thought back on that summer, I realised that death is not always
a cruel, unseeing killer. Sometimes it leaves the door open on the
way out.

But being set free can be sheer torture. The day it happened
was not the usual day of hard work and evening sun, a day that
Furtwängler's baton slowly lays to rest. The fact is, the day before
Bestefar died someone painted a swastika on his car.

I had been waiting all week for a delivery from Oslo, and now
the delivery slip was in the postbox. I took the shortcut to the
house, racing past the stinging nettles and across the farmyard.
I opened the door to the tool shed just a crack and said I had to
pick up something, that I was leaving now. He straightened up at
the workbench, set down the pair of pincers and said we should
drop in at the agricultural cooperative.

"Why don't we take the Star?" he said, and brushed the wood
chips off his jacket. "Save you money on petrol."

I turned and closed my eyes. So it was one of those days, when
he thought it embarrassing if we drove in separate cars.

Bestefar ambled stiffly across the yard to fetch his shopping
jacket. The villagers did not like it that he carried a knife, so he
usually wore a hip-length jacket when we drove in.

We set off in the heavy, black Mercedes he bought new in
1965. The paint was scratched from driving along the overgrown

pasture road and there were patches of rust around the lock for the boot, but it was still a car that stood out in the village. We drove slowly past the potato fields, inspecting the blossoms on either side.

We were potato farmers, Bestefar and I. Yes, we had sheep, but potato farmers was what we were. He would lose weight waiting for them to sprout, even though the fields of Hirifjell were 540 metres above sea level and insects that spread disease rarely made it this high.

Bestefar was one hell of a potato farmer, and he made one out of me, too. We supplied both seed potatoes and eating potatoes. Almond potatoes were the most lucrative, even though Ringerike were a better quality. Beate was a potato for idiots. Large and fla-vourless, but people had to have them. Pimpernels were what we kept for ourselves, for meal after meal. They matured late but had a firm flesh, and with their bright violet skin there was nothing more beautiful to pluck out of the soil.

The car rattled as we drove across the cattle grid, and he turned onto the county road without checking for traffic. The woods opened near the Lindstad farm and as always we studied the river.

"Laugen has gone down," he said. "We could go fishing past the campsite."

"Graylings won't bite when the water is this green," I said.

The spruce trees closed around us and the river disappeared until we reached the asphalt. We rumbled down the steep hill and I felt a flutter in my stomach, as I always did when approaching Saksum. The train station, the secondary school, the sawmill, the barns on the sunny side. The people.

Cold river air poured in through the window as we crossed the wooden bridge.

"Agricultural co-op first?" he said.

If he went in there, he would be some time. Bestefar did not go in for window-shopping, so we rarely left without a tail-heavy Mercedes and a receipt half a metre long.

14

"As a matter of fact," he said. "Why don't we pick up your package first. Let's do that."

We had just come out of the post office and I was examining my brown cardboard box when I noticed a strange falter in Bestefar's step. When I looked up I saw the clumsy swastika sprayed in red paint on the door of his Mercedes.

That had been my exact thought. *His* Mercedes, now that it had a swastika. Earlier that day, and in all the years prior, I had considered the Star *ours*.

People were staring. They stood by the noticeboard for the athletics club with their hands in their pockets. Børre Teigen and his old lady. The Bøygard daughters. Jenny Sveen and the Hafstad boys. They were staring at something behind us, as if the post office had new roof tiles.

Thin streaks of paint trickled slowly down the car door.

One of the Hafstad boys glanced up at the corner by the draper's. The flapping of a coat tail, someone slipping away. The only movement in a few frozen seconds on this Saturday in Saksum.

Bestefar lowered his arm in front of me like a boom.

At that point I still had a choice. The village was watching and waiting. Again I chose Bestefar. As I always had, not lacking for opportunities. I broke past his arm, dropped the package and I was off, running as I had my entire life. Swiftly past the glare of the villagers, across the road, towards the gravel playing field behind the Esso garage. I caught up with him there. A teenager running awkwardly with his arms pinned to his sides, his grey nylon jacket fluttering behind him.

Obviously I should have used my advantage, my speed, to pull ahead and stop him face to face. I should have taken him with my size, slowed up like a footballer after scoring.

But instead I stretched out a leg and sent him sprawling head-first into the gravel. He screamed as he fell and kept screaming when he was on the ground. I grabbed him by the jacket and spun him around.

15

Noddy.

In fact his name was Jan Børgum, but he nodded incessantly and talked to himself. There was dirt in the grazes on his hands and sand in his hair. Tears flowed into his bleeding nose, and pink bubbles spluttered as he sobbed. He had spray paint on his fingers and on the sleeve of his jacket, and he was clutching a piece of wax paper with the clumsy outline of a swastika.

I cursed to myself.

"Jan," I said. "Did somebody pay you to do this?"

He gurgled something unintelligible.

"Speak clearly, Jan."

But he could not speak clearly. I knew that.

I tried to help him up. He pulled away, fell on his backside and cried even more loudly. His trousers were torn at the knee. A pair of grey trousers, the kind old people and taxi drivers wore. Jan's mum had been dressing him all these years. He had been two years above me in primary school and had worn the same clothes back then, when he wandered around with the special needs teacher, cross-eyed and his mouth agape. When I began secondary school, Jan did not progress to Year Nine. Jan went somewhere else.

People were coming. They gathered by the oil-change ramp at the Esso.

"Come on, Jan," I said. "Get up."

He sniffled and wiped the blood off his lip. Struggled to his feet. I asked if he was hurt and he nodded. I gave him some money for the ripped trousers. "Who told you to do it?" I said.

"It was in the book," he said.

"What book?"

He mumbled something.

"If they come back, tell them I want to talk to them. Can you do that?"

"Do what?"

I brushed the dirt off his back. He stood gawping. I headed towards the Esso. The Bøygard girls turned away. Then the Hafstad boys, and finally the rest of the crowd broke up. They strolled back to their shopping and their cars and their hot refill coffee.

If only they would attack me, lay hands on me, yell at me. Then I could have responded, taken up this argument in the centre of Saksum, in the middle of a shopping day.

But *how* would I have responded? Besides, they were done with their staring – at the rabble who settled things amongst themselves – and now that everything was over, they had two fewer halfwits to worry about.

Bestefar was in the passenger seat. He said nothing, did not wind down the window. He just sat there like a wax figure in a German fighter plane, pointing at the steering wheel. He had not touched the spray paint. You did not have to be a fortune teller to know that Sverre Hirifjell would never give anyone the satisfaction of asking for a rag. Or go into the paint shop to buy Lynol while maintaining an appropriate level of anger, mumbling something about juvenile pranks. I am not even sure he knew what that meant.

I opened the car door. People at the general store were taking their time to park.

"We are not driving like this," I said. "I'm going to wash it off. Or cover it up."

"Drive," he mumbled. "Straight to Hirifjell."

My package was on the back seat. One corner was crumpled.

"Just get in and drive, dammit," he growled. "Straight through the village. Up to Hirifjell."

He offered no objections when I took another route. I drove down to the grain silo and took the gravel road along the river. Six kilometres longer, but nobody lived there, and the swastika faced towards the mountainside.

"It was Noddy," I said.

But he just stared at the river, and I realised that he was fully occupied with the one thing he truly was good at: forcing himself to forget.

17

*

The sky behind the barn had grown dark. I strolled across the farmyard and sat on the front steps of the cottage. The Mercedes was under the barn ramp. Bestefar was inside the log house.

I have never liked seeing people mope. Almost anything can be fixed. Tobacco and coffee help. That, and putting my cards on the table. If you have a two of clubs and a three of diamonds, then fine. Today, you lost. The only grounds for complaint is if you are dealt four cards when you should have had five.

There was rain in the air and I wanted it to come. Rolling down from the hillside with a strong wind. I wanted this rain, I wanted to put on some coffee, go out to the conservatory and hear the raindrops drumming on a roof I had built with my own hands, while I sat nice and dry with my mug and a cigarette.

I went to the storehouse and pulled a tarpaulin over the circular saw. That week I had changed the gable board and rafters, now all that remained was the painting and I could do that after the weekend.

The rain was close. Good rain. I recognised its smell. Not too hard or heavy, but one that would last a long time and give the ground a good soaking. I had been planning to take the irrigator out to the north field this evening, now there was no need. Instead I kicked off my shoes and put on a pair of thick, woollen socks. While the coffee machine gurgled, I cleared the kitchen table. Wiped it down with a cloth and reached for the package.

Oslo Camera Service knew their trade so well that their reputation had travelled all the way to Saksum. Tightly wrapped, brown-taped corners. My name typed out, themed postage stamps, C.O.D. slip filled out with no abbreviations.

I cut open the package, found another box inside and pulled out a lens wrapped in tissue paper.

Leica Elmarit 21mm. A wide-angle.

The weight of it. The resistance of the focus ring. The inscrutable colour shifts on the glass coating. The silky matt lacquer, the engraved numbers indicating distance and aperture.

Bestefar gave me the Leica for my eighteenth birthday. An M6 camera body, a Summicron lens and ten rolls of film. There was no better camera for miles around, unless someone owned a Hasselblad. The only thing that spoiled it for him was that the lens had distance markings in both metres and feet.

"There's no need for that," he said. "No enlightened nation measures in feet."

I bought myself a new lens every year, for a sum most people would consider too steep for a television. The world was made new with every focal length. A telephoto lens that brought the subjects closer and left the small detail in a haze. The macro lens that made the corolla of a flower fill an entire planet. And now, a wide-angle lens to expand the horizon outwards, making medium seem small, and bagatelles into a speck in the eye. It demanded a different subject matter, new ideas about foreground and background.

But that day I did not look through the Leica's viewfinder, because I would see only the usual. My collection of *Asterix* books. The door to the darkroom. The stereo with the speakers I built. The glass cabinet with the rest of my camera equipment. A still of Joe Strummer from the filming of "Straight to Hell". The huge poster of The Alarm, from the cover of "68 Guns" where nobody looks into the camera. On the wall, a long line of my nature photographs.

I knew where I should go to take photographs: the flame-birch woods. But not until early the next morning.

I moved into the cottage when I was sixteen. The house had been empty since I lived here with Mamma and Pappa. Back then, I kicked open the swollen door without thinking that now, now something historic is happening. I just began to use it, put up new panelling inside and built a conservatory from which I could see the edge of the forest.

The house was mine, and at the same time it was ours.

A little of the two of them remained. The Mixmaster, Pappa's wellies, the bedclothes. I left the photograph of the three of us in the log house. I still felt that I should stop each time I passed it.

When I was younger, the photograph was a hope. A hope that Mamma and Pappa might not be dead after all. Later it became a reminder that they were never going to call. For a long time I wondered why Bestefar had placed it by the telephone instead of hanging it on the wall. Was it in order to remember them, or was it so that the picture would have an effect on us when we spoke to others? Or to remind us that those who phoned *here* also had Mamma and Pappa's story in mind when they expressed themselves?

My grandmother's name was Alma, and I never called her anything else. She was quiet and guarded, like an old floor clock. An illness left her bedridden, she moved to Kløverhagen nursing home and was laid to rest when I was twelve.

But now and again she shared small details about Mamma. She told me that her family had been killed during the war. That was why the issue of adoption rights never arose, that was why they never expected any French relatives to visit. She talked about her little, but that did not surprise me. Because my father's side was not big either, only a few second cousins. We never went on trips, just to the occasional funeral, and we would always leave before coffee was served.

All the same I was surprised; even if Mamma's family was gone, surely not *everyone* around her could have disappeared?

Those were my thoughts when the two of them took an afternoon nap on their respective couches, and I opened the atlas and studied France. Told myself that *somewhere* there must be *someone* who remembered Mamma, because she had lived for almost twenty-seven years. I looked for Authuille and read about the Somme and the First World War in Bestefar's encyclopaedia. Imagined a village and a war.

Every so often we went to the churchyard. The smell of tar from the stave church followed me to a gravestone made of blue Saksum granite. WALTER HIRIFJELL. NICOLE DAIREAUX.

Mamma born in January 1945, Pappa in 1944. Both died on September 23, 1971.

But I turned back before I stepped too close. When I asked myself how Mamma and Pappa met, I curbed my curiosity. I didn't want to allow them to appear before me. You cannot miss something you haven't had, I told myself. Bare ground must not lie open; all black earth was a wound. It attracted weeds which grew and covered it.

Even now, here in the cottage, they occasionally stepped out from the shadows. Once I found an L.P. of French songs for children, and when I put it on, there was a glimpse of my mother.

I knew all the songs. I had sung "*Frère Jacques*" instead of "*Fader Jakob*". And I understood the lyrics to "*Au clair de la lune*" and "*Ah, vous dirai-je maman*". The busy language came easily to me, and I realised that I must have spoken French when I was little. Mamma had sung with me, our voices had filled this house.

French was my mother tongue, not Norwegian.

At secondary school, it was either German or French as an option. It was the first time I felt I had to choose between my parents and Bestefar, and I kept it from him that I chose French. Mamma's language was awakened within me, so rapidly that my teacher wondered if I was messing her about.

Later I found more traces of them, in a large cardboard box in the loft. A make-up bag, a razor, a wristwatch. The way the belongings were thrown in together told me that it had been painful to sort through them.

At the very bottom there lay a book. *L'Étranger* by Albert Camus. I flipped through its pages, studied the sentences, pictured Mamma sitting there reading. Then I got a shock, followed by an expectation. Like seeing a fish leaping, out of casting range. On the first blank page, written in blue ink, the words *Thérèse Maurel, Reims*. They must have been friends. Once their hands had held this book, at the same time, or almost the same time.

I was no longer the only proof that Mamma had existed.

*

I began that day to form a plan, to visit the place where Mamma and Pappa died, see if it would awaken something in my memory. Because there had been an eyewitness: me. It must be somewhere in my memory, like photographic emulsion that has once been exposed to light.

Sometimes it ached inside me, the urge to leave. But the world stopped in Lillehammer. South of Helge Menkerds Motorsenter, everything was alien, I had no experience of travelling and I had no explanation for Bestefar, his eyes would take on that familiar wounded look; was he not enough for me, had he not done everything he could?

As a young boy, it was I who needed Bestefar and the farm that needed him. Then I got older and was given my share of the work at Hirifjell, and soon the farm and the sheep needed *me*. The longer I waited, the older he got, and when I was around twenty, these needs converged, so that it was just as difficult to leave as it was to remain, and from that day everything settled in the path I took, a path which gradually grew deeper and more habitual.

It came off with Lynol.

The swastika dissolved and disappeared into the rag. Pink waste water looking like something infectious. I got dizzy but moistened another cloth. Picked some grit from the enamel and rubbed harder. The fumes, thinner than air, crept into my lungs. I dropped the rag and ran out into the rain, stood looking at the Star under the barn ramp. The outline of the swastika was still visible.

I returned to the stench of Lynol. Rubbed and rubbed. Somewhat woozy, I walked across the farmyard, up the stone steps and into the log house.

"Managed to get it off," I shouted.

No answer.

The cuckoo clock showed half past four. I could tell by the smell of tobacco that he had been standing by the door. I went up the stairs, stopped halfway. Heard his steps on the second floor. What kind of fresh incursion was this? We never used the rooms up there, they were cold and dusty. I stood by the map that marked out our parcels of woodland.

"Driving into the village," I said, as though speaking to the stairs.

His footsteps paused. Then he shuffled on.

The town centre was dead. I knew it would be, nobody was out in the languid hours between closing time and suppertime. Nothing more than through traffic winding past at fifty. They glanced out of their car windows, pleased not to be living in Saksum.

But they did not know what we had.

Because there was room for us here. Room for me, for Carl Brænd, the electronics freak who lived with his mother at the age of fifty-five, who built ingenious amplifiers and drove to the convenience store five minutes before closing time to buy anaemic hot dogs at half price.

Our shortcomings were visible here. We knew about them, used them to torment each other, but local gossip kept us together. There was a hole in each of us, and we searched for the hole in our self-righteousness because that was the common thread that held the village together.

I drove round the town centre and back down to the Salvation Army, seeing nothing more interesting than my old moped outside the Norol station and two children running up from the football pitch. I drove back to Laugen. Rolled down the window as I passed the secondary school, noticed that the air was getting cooler.

I heard the roar, saw the water. Inside the glove compartment, I found the Bob Dylan cassette Hanne had left. "Knocked Out Loaded". It had disappointed both of us, apart from "Brownsville Girl". I played it anyway. She was in the village, and when the song came on it was O.K. to admit that I was driving around looking for her. A few days earlier I had seen her outside the draper's. Wearing a light-brown suede jacket. Like an antelope, with her chestnut-brown hair and her long legs.

Her presence, so typical. She must have seen me first and then slipped into the clothes shop where she knew I could not follow her because I was wearing grimy work clothes. One second we were looking at each other. The next she was gone.

Hanne was one of those girls who was grown up from the day she was born. When she was fourteen, she borrowed her brother's moped without asking and drove out to see me. She stood by the postbox flashing her lights. Like a smuggler on the shore signalling to a fully loaded ship in the middle of the night.

We slept together long before the official threshold, but gradually she gave me the feeling that she had to *rescue* me. That I was the mangy puppy she was bringing home. She harped on about the word I despised: "education", this compulsory ski run that passed through Oslo or Bergen or Ås, as though we were all obliged to collect something and return with it to the village so that it did not fail. I did not want to be filled up like a thermos. As I saw it, I had no obligation to anything or anyone. Apart from travelling to France. But when I said that to Hanne, she countered with a "Why?"

"Let it go," she said. "You came back unscathed. You'll find nothing but old memories to torment you. What can *you* uncover, nearly twenty years later, that professional investigators could not find at the time?"

It irritated me, her choice of words. *Professional investigators*. As if she were reading it aloud from a book. She blocked my path like a white picket fence, but still, I did not go to France when we broke up. I just started up the tractor and drove out to the fields.

The years had passed, but her telephone number still lived in my fingertips. Eighty-four for Saksum, then her number, diagonally opposite on the keypad. She would hear about it this evening at a party. Someone uncapping bottles of Ringnes and dropping hints about me. Girls gathered together on the couch, perfumed and half-drunk, sideways glances whenever my name was mentioned, the guy who made a fool of himself in the town centre, what do we think about him, would anyone defend him, *can* anyone defend him?

Yngve's Ford Taunus appeared. He flashed his lights and we pulled up alongside each other by the fire station. I rolled down the window and found myself looking around, yes, *hoping* someone would see me with the chemist's son. The guy who left sixth form with so many top grades that people called him Maxi Yatzy. While I dropped out of secondary school and that was that.

"Laugen is going down," he said.

I had always liked that, sitting car to car around five o'clock on a Saturday, the raised rear end of a blue Opel Commodore GS/E next to the 20M chrome grille of a Ford Taunus, glistening after two tubes of Autosol. As long as there were locals in the village, five o'clock was a nice idle time. A time that did not differentiate between those who worked and those who went to school, a time when the only difference between us was that he smoked Marlboros and I smoked rollies. Yngve had been going out with a gorgeous girl from Fåvang called Sigrun, but had recently dumped her because she was "too fussy".

"Sigrun was not fussy," I said.

"No, but that's how it was," he said.

We were quiet for a moment.

"It's just a bit odd," I said. "Like not liking Bruce Springsteen."

"I don't like Bruce Springsteen," he said.

We discussed whether it would be best to fish at the mouth of the river with rods, or if we ought to prepare for a longer spell on a boat with otter boards. I did not ask if he was going to the party later. I assumed he was. Yngve was the type to arrive late and draw a crowd.

"Seven o'clock, then," I said and looked at the clock on the dashboard. "Just have to get some food in me."

But he did not roll up the window.

"Heard there was a bit of a commotion in these parts," he said, nodding towards the post office.

"Commotion?" I said. "All hell broke loose."

He looked down at the car door and tapped his cigarette.

"What are people saying?" I asked.

"Just that he sprayed the car and you got angry."

"Bah. That nasty Hirifjell kid beat up poor Noddy, that's what they're saying."

"You didn't beat him up."

"How do you know that?"

"People are saying that you did *not* beat him up. That you stopped when you saw that it was him. Brushed him off and let him go. That's what people are saying."

I took one last drag and dropped the cigarette between the two cars.

"People know," Yngve said. "People know who he is. That he spends his time at the day centre. That he gets up to things like that."

"Meet you by the river, then," I said. "We'll go otter boarding."

The water for the potatoes came to a boil. I took the pot off the hob, dropped in a spoonful of rock salt and grabbed some medium-sized pimpernels. A few extra for tomorrow's breakfast. Always the same, fried potatoes with herb salt, salted pork and three eggs each. That would keep us going until the newspaper arrived, even if it came late.

In the living room, Bestefar was snoring on the couch with his feet up on a yellowed edition of the *Lillehammer Observer*. His Russian bayonet on the table. A dead cigarillo in the crystal ashtray. He must have dozed off before he finished smoking it.

I took the tartan blanket from the television chair and spread it over him, checked the pill box in the chest of drawers to see if he had taken his medicine. I went to the kitchen and took out some Wiener schnitzels, then fetched sugar-snap peas and lettuce from the garden. Parboiled the peas and set the table. When I shouted into the living room that dinner was ready he did not wake up. Fine by me. Conversation was not going to solve anything anyway. I finished eating and got up with my mouth full. I slammed the hall door to wake him.

I awoke with the Leica on my lap. The morning light was nearing. I was on the far side of the sun.

This was my hour. Leica hour.

I went outside. The smell of raw grass after rain. A magpie took off from the nettle where I had thrown the fish guts the previous night. We had spent four hours otter boarding on Laugen, close to the tall, dark mountainside the trout kept in the shadow of, then out into the current where the graylings might bite at any time. We had a good laugh and drank Coke, smoked and chatted in the blue exhaust of the Evinrude outboard, stopping only when the otter line twitched in our numb hands. At home I scrubbed my fingers under the tap until they tingled, sat down with the Leica and fell asleep.

Now I walked across the potato fields. Below, the farmyard appeared through the mist. The light outside the log house was on, I looked down at the barn and the tool sheds. Then I went on, into the flame-birches.

As a child I had been terrified of coming up here. I would hear loud bangs in the spring, like someone firing a shotgun. Bestefar heard the same, would straighten up and look towards the woods.

"That's my brother's iron snapping," he said and went back to what he was doing.

Never before had I heard him say the word "brother". Later, I discovered that his name was Einar, and that there was hostility between them. They were on opposite sides during the war. Bestefar went to the Eastern Front and Einar went to Shetland. Little more was said, just trivial things that Alma mentioned, like when the living-room table got scratched. "Oh, it's only something Einar made", was how she put it, and when I asked, Alma said that he was a cabinetmaker who had worked in Paris in the thirties and had been killed in 1944.

He left behind a workshop. The building was a little isolated, a flaking, red, elongated cottage. The insides of the windows were covered in dust, and it was the only building around which the weeds grew wild. But back then, when this was all new to me, I did not ask about Einar, only what Bestefar meant by *iron*.

"My brother put iron bands around the trees," he said. "They're rusty now. At this time of year, the sap rises. The trees grow. The cracks you hear, that's the birch breaking free."

I could not comprehend why Einar would torment the trees.

"Keep clear of those woods," said Bestefar. "Splinters can fly off. And if there's one thing you don't want to see, it's flying iron splinters."

Then he got that rare look, one that would startle me and simultaneously drain me of compassion, and I knew he had drifted back to the war. Often he would have second thoughts, so that it was followed by a strange, gentle and somewhat uncertain expression, and then he would climb down from the tractor and ask what I wanted for supper.

I said I was happy the trees could not shout when they were in pain, otherwise I would never get any sleep, since my bedroom window faced an entire screaming forest. But I said that only to make Bestefar happy. I never even asked why Einar had fastened iron bands around the trees.

Then I read *Det Hendte 1971*, and in the long hours when I was angry without knowing why, it was as if I had taken Einar's side because he had fallen out with Bestefar. Right after the first thunderstorm, the time of year when the sap begins to flow, I lay waiting for the cracks from the birch woods. And one night it was as if I wanted to see Einar. I crawled out of bed, sneaked past Bestefar's bedroom and put on some clothes I had hidden down in the hall. Ran towards the woods, glancing back to see if a light came on in the window.

The ground was damp from the rain. The moon was large

and cast long shadows where I walked. High on the hillside I glimpsed the spruce branches of the surrounding woods broken by green foliage. I crouched as I drew nearer. The undergrowth was thick and brushed me with dew.

Then I stood amongst the birch trunks. He had fastened some kind of withies around all of the trees. Flat, rusted iron bands braced against the white birchbark, while a sea of green leaves rustled in the treetops high above. It was a large parcel of woodland and there must have been a hundred birches with rings. Five or six on each tree, at various heights. He would have used ladders to attach them. The bands were meant to be adjusted as the trees grew, as there were long bolts with huge wing nuts on the ends. But he was shot in 1944 and never returned to loosen them. Most had rusted apart and hung loosely from the trunks, a few were ingrown into the tree, while others had fallen off and were poking up from the forest floor.

Why did he torment the trees? I stood there for a long time that night, between white trunks that seemed to be an infinity of flagpoles, rehearsing an anger towards a man who was dead, an anger which I soon set aside because I realised that I was merely copying Bestefar.

There was a crack behind me. I raced back down to the farmyard along the same path I had made on the way into the woods. I crawled back under my duvet and lay there, breathing rapidly, and then I did something I had not done in years: I slipped into Bestefar's bedroom and lay down in his closet, staring up at the shirts and trousers on the hangers.

I was terrified, truly terrified. The crack in the woods had woken something inside me, a pervasive fear and a memory which stirred deep within. I thought I heard voices in the distance. Then, in the midst of all that obscurity and menace, a memory of a toy dog so distinct that I wondered if I had simply imagined it. It was made of wood, had drooping ears and could nod its head and wag its tail.

But was the memory real or just a wish I comforted myself

with? I had never owned a toy dog. Maybe it belonged to someone we visited, and I had confused the memories. Because even we must have been places. Visited people.

Been normal before we died.

The following day I asked the woodwork teacher a question. He swept the wood shavings from his leather apron and said: "Flamebirch? The finest cabinetmaking material in the country. Comes from trees that are scarred in some way. The pattern emerges from the tree doctoring itself."

That was his expression. Doctoring.

I had never heard the woodwork teacher talk that way. Usually he would go on about the importance of not wasting material, about making more precise measurements. Now he disappeared into a closet and came back carrying a small cupboard door which had a golden shimmer. The meandering pattern created shades of black and shadow play on the luminous, amber-yellow woodwork.

"What you see are scars," he said. "The tree has to encapsulate the wound and continue to grow. The growth rings find alternative routes, extend across the wound. The pattern is unpredictable. Only when you saw parts off the tree can you see how it will turn out."

I was good at woodwork, could make seamless joints and carve small faces freehand. "Cabinetmaking is in your blood," he said thoughtfully, and I felt a tugging inside me, a connection that did not end at Hirifjell, but beyond.

I was forever returning to the woods, but never told anyone that I went. I would sit there looking at Einar's trees in shackles. It became our place, Einar's and mine, and when I had a row with Bestefar, I was quick to think of Einar. I would imagine him coming down from the flame-birch woods to argue my case. I sat up there watching the birds, listening to the rustling of the leaves. Dreamed up explanations as to what might have happened in France. That Mamma and Pappa were alive down there. That I

30

had been swapped with another child and brought back here. That I was infected with a dangerous illness, which Mamma and Pappa did not have the strength to see in full bloom.

Later I killed off my fabrications one by one. Over the passing years there were fewer cracks, most of the withies surrendered to the trees and ruptured, and the false visions gradually disappeared.

Bestefar shied away from that parcel of forest. It would have been natural for him to fell the trees for firewood, thin the woods and keep them tidy, but he never went near them, and he made it clear that he didn't like the idea of me going in there with a saw either.

But then something happened that I could not explain until many years later. The night before my tenth birthday, I was woken by a noise and I got out of bed and went down the hall. I heard Bestefar downstairs; he was angry and said something I did not grasp word for word: *I do not want to be haunted by that*, or, *spare him from being haunted by that*. The rest boiled over into an outburst filled with hate, and when I heard his step on the stairs I hurried back to my room.

From the window I saw a car I did not recognise, heard voices and the humming of the engine. Then the car turned, its tail lamps leaving red streaks in the darkness as it left. The next morning, Bestefar said that some travellers had shown up at the door, bothering them for directions at an absurd time of night.

On the kitchen table there was a layer cake and enough food for two days up at the mountain pasture. It was supposed to be a surprise, we were going to celebrate my birthday.

But on the drive there, I had the feeling that Bestefar and Alma were afraid of letting something slip, and that night I dreamed that I was surrounded by a crowd of people who were laughing at something written on my back, but I was unable to take off my jacket to see what it was.

A few days after my birthday I took one of my usual walks up to the woods. But when I stood amongst the tree trunks, the place felt disturbed, haunted almost. Then I saw the stumps. Four trees

31

had been felled and limbed. The sawdust was yellow and fresh, the shorn edges were bleeding sap and there were flies buzzing around.

I got down on one knee and let the sawdust trickle through my hands. Large, round grains, from a coarse-toothed bow saw. The limbed branches formed silhouettes of the trunks, and judging by the distance between the piles of chips, I saw that the trees were cross-cut into two-metre lengths. In the grass I could see the trails the trunks had left; they had been dragged over to a slope, then rolled down to the county road. This was not illegal felling for firewood, because there were trees closer to the road. Whoever had been there had known what he was looking for.

Now I stood amongst the tall, white pillars of birch with rusted rings, this time gripping the Leica. Some had broken free of their shackles since I had been here last, others had succumbed in their battle with the withies and allowed them to become embedded. I shifted my position, studied the direction of the shadows, my eyes searching until they found the theme.

The sun arrived. I lay on my back and looked up. Through the wide-angle lens I saw the trunks reaching for the heavens. That would be a good one. I saw exactly what I wanted to see. The foliage, the clouds, the trunks and the foreign element – the iron – which would make this a photograph and not a picture.

The shutter emitted its brief whisper, the Leica capturing something in the *now* and allowing it to become something from the past.

As I stood up I pricked my finger on a jagged iron withy. I sucked off the drop of blood and walked back to Hirifjell.

He was not at the kitchen table. That was the first thing I noticed.

Because Bestefar should be sitting right there, wearing his dark-blue work jumper, fried eggs on the hob, two coffee cups on the table, glancing up from yesterday's edition of the *Lillehammer*

Observer. He should be sitting right there, as steady as the log walls behind him, folding up the newspaper when I came in.

But the table was still set for supper. The water in the jug was cloudy with air bubbles. The peas in the bowl were shrivelled. In the frying pan there were two dry Wiener schnitzels.

I walked slowly into the living room.

He had the same tartan blanket over him, his feet up on the newspaper. I stopped in the middle of the room and thought, *Now it begins*.

Bestefar was lying on the couch, and Bestefar was not sleeping.

2

I HAD THOUGHT HE MUST BE ALREADY IN HIS GRAVE, OR at least no longer able to drive. But it *was* him, Magnus Thallaug, the old priest, driving the matt, dark-blue Rover I remembered from the time of my confirmation studies. The car wound its way down from the gate and rattled over the cattle grid.

I tucked my shirt in. Ran my fingers through my hair.

The priest was peering through the grubby windscreen, both hands on the steering wheel. The Rover stopped in the middle of the farmyard, where the hearse had stood. The door opened, the priest tested the ground with his walking stick, set down one skinny leg. The patch of skin between his socks and shabby suit trousers glistened, pale as skimmed milk. He hoisted himself out of the car and looked around.

"You have to eat, Edvard," he said when his eyes came to rest on me, as though he had checked on the status of the barn and storehouse in turn. "Otherwise there will be no farmers left at Hirifjell."

Hesitantly I shook his hand. His skin looked two sizes too big. When he opened the back door, the smell of a hot old car seeped out. On the cracked leather seat there was a threadbare bible, some pages of which had come loose.

"Letter to the Ephesians," he mumbled and shuffled the pages back in place. "Been like that since the New Year's sermon in 1956 when The Word fell to the ground in front of Reidun Ellingsen. She was sitting on the front pew half asleep. She has been devout ever since."

"That was probably for the best," I said.

"Absolutely. Listen, Edvard, the fact of the matter is, I've taken on some summer work. The Faculty of Theology produces people who require holidays these days, you see."

He shifted the bible to his other hand. "In my time I worked all year round. True, it was with godless peasants or empty pews, but I was there."

"Yes, of course," I said, and knew that I qualified for both lists.

"And now I have to officiate at the funeral of Sverre Hirifjell."

I stared across the fields.

"Listen, I realise you're in your own world. But we have to sit down to plan your grandfather's funeral. And as I said, you need to eat."

"Let's do what has to be done," I said.

It sounded so easy when I put it like that. But it had not been easy earlier that day. I had been looking at the wall clock tick away for what must have been fifteen minutes. And at Bestefar, at the Russian bayonet in its sheath on the table, at the aerial photograph above the couch, the photograph of our farm, which was now mine.

Then I did something I did not expect of myself. I fetched the Leica and with trembling hands I took a photograph of my dead grandfather.

Where he lay.

Just as he was.

His lips forming an expression I had never seen while he lived. His dry eyes. Him, and yet *not* him. Like a statue of himself and his life.

Later I rang the authorities and the Landstad funeral parlour and went back downstairs. I stood motionless, holding the Leica, thinking that in there, inside the camera, he was less dead.

Only then did I notice that the Grundig amplifier was still on. The first act of Wagner's "Parsifal" was on the record player.

He had always looked at me strangely when I asked him to play it. I moved the needle to the first groove and the music began to soar, and I stood like that and he lay like that, until I realised that there were people around us.

I went into the hall and heard them talking. It sounded like the police chief was vying to outdo the doctor in competency. They mentioned a stroke, and then an hour or two passed without me knowing whether or not they had left, until Rannveig Landstad and her son stood there. For three generations the Landstads had run the funeral parlour in Saksum, and because the son was one metre sixty tall and destined for the same fate, he was known as the Mini Digger. I had called him that once at a party, but today, in so far as his errand concerned me, the nickname seemed cheap.

They just took him. Wearing the clothes he died in. Carried him off on a stretcher down the stone steps into the hearse. I felt they worked too quickly. This was Saksum, there was no risk of another parlour offering better work for a lower price.

Then they came back inside and spoke of "support" and "the grave occasion", and they were in no hurry to leave until I was close to my old self again.

"How do we proceed?" I said.

"Of course the coffin is already taken care of," Landstad Junior said, as if keen to demonstrate his standing, but Rannveig glowered at him and he went quiet. "Come by when you are up to it," she said. "We can take it from there."

I looked at the couch on which Bestefar should be lying. "Did he know that he was going to die soon?" I said.

She knitted her brows.

"Seeing as he had selected a coffin?" I said.

Rannveig Landstad was about to say something. Exchanged glances with her son, and for a fraction of a second I thought I detected annoyance. Then she shook her head. "Come down whenever it suits you," she said. "We have to take one thing at a time."

I let it go. They went outside and switched on the cross on the roof of the car. "Wait," I shouted. I raced into the living room and fetched the Russian bayonet, opened the back door and climbed in with him. The light shone through the pale-yellow curtains and made his face appear healthier, and it was as though he was returning to me. I opened his buckle and eased the sheath with the Russian bayonet onto his belt.

"You never got so old that you needed help getting dressed," I said quietly and positioned the buckle in the dark depression in the leather thinking that now, now Bestefar could finally go down to the village without anyone turning their nose up at him for carrying a knife, and then I whispered "Good night, Bestefar", so quietly that I did not hear it myself.

All of this boiled inside me and I must have been more in my own world than I realised, because the old priest took me by the shoulder and said loudly: "Now listen: there or there?" and pointed with his bible, first in the direction of the log house, then at the cottage.

The log house it was. He made straight for the kitchen. "Everything appears to be as before," he said, and gazed at the corner cupboard with the stuffed wood grouse, at the blue pantry door, at the woodburner. Pulled out a stool and sat down at the end of the dining table. Presumably he had experience of this, of avoiding sitting somewhere that might have been the seat of the deceased.

"I haven't had a chance to clear the table," I said, and began to tidy up.

"No, wait," he said and placed his walking stick over my wrist. "That plate there. Was that for Sverre?"

We do not set a place for the cat, I thought.

"You made dinner for him yesterday. Which he didn't get a chance to eat."

"I made dinner every day. For both of us."

"Listen, Edvard. I understand it was a stroke. That, what should we call it – that *incident* in town yesterday. With the swastika. Does the police chief know about it?"

"The police chief in Saksum knows everything," I said.

"And? Was there a connection?"

"Noddy doesn't know what he's doing. There's no pinning the blame on him. People have harassed Bestefar with swastikas before."

"Hm," the priest said. "Sit down and eat, Edvard. Take his plate. Do not let a work of creation go to waste. Particularly not Sverre Hirifjell's last meal."

I reheated Bestefar's Wiener schnitzel and made coffee. The priest pulled out a handkerchief with thin, violet stripes, blew his nose and said:

"There has to be music when people come in. But the organist needs to be restrained a little. He's come straight from the conservatoire. Has no idea that a funeral must have flair."

Thallaug planted his cane on the floor and pottered into the living room, towards the music shelf. Put on his glasses and searched through the most worn-looking records. "Bach's trio sonatas," he said, bent over, "while people are finding their seats. Then something with a little sizzle."

He pulled out an L.P. and ran his index finger along the titles. "Perhaps Buxtehude? We can hardly expect a great crowd, so we might as well choose something that fully captures Sverre's spirit."

He probably hasn't considered that I have to be there, I thought. I have to endure the music too.

"What about '*Maurerische Trauermusik*'?" I said. "That's a good one."

"Mozart?" he said from the living room.

"Yes. Can we use that even though he wasn't a Freemason?"

"Absolutely. We're making progress."

"Sverre knew his music," the priest said while I chewed on the crusty schnitzels. "Hmm, all those unsuccessful organ concerts in Saksum church over the years. Hardly a soul to be seen. We could have put Peter Hurford in the programme and no-one would have known who he was. But your grandfather was a fixture. Always found the place with the best acoustics. Fourth row of pews, close to the nave. Was never seen in church otherwise. In fact he was just as artistic as his brother. Oh yes, a fine piece of music sometimes brings people closer to God than any priest can manage. We are many who speak of heaven. But few who understand eternity."

I fetched the coffee pot. "When did you come to the village? If you don't mind me asking."

He did not answer at first. His eyes were travelling. Searching the log walls, looking out of the window.

"I arrived in 1927," he said. "For fifty-five years I have served the congregation in Saksum. I married Sverre and Alma. I baptised, confirmed and – unfortunately – buried your father. I buried your mother beside him. I baptised and confirmed you. But I assume you and your grandfather have let what happened to your parents rest in peace."

I looked down at the table. He seemed to be taking stock of me.

"He had wanted to go otter boarding in the evening," I said. "I should have tried to wake him."

"Edvard. Do not punish yourself with thoughts of what you could have done differently. If you look at life as a whole, most of our conduct is second-rate. We are blind to the goodness people are prepared to offer us. We only half listen when someone tells us something they have dreaded saying. Death does not send us a letter giving three weeks' notice. It arrives when you are eating raspberry sweets. When you have to go out and mow the lawn. Now it has been here too, helping itself. But you can find comfort in the knowledge that it will be a long time before you see it again. That is why, after the funeral, I would like to have a chat with you about your parents."

"About Mamma and Pappa?"

"Yes. Whenever it's convenient."

"I think I'm too poorly dressed for a serious conversation," I said. "But we might as well do it now. Then I can take the whole load at once."

"No, we should wait."

"I won't cry," I said. "What did you want to tell me?"

"Well, how much do you actually know about them?"

He had eyes that were impossible to lie to. I shrugged.

"It's mostly a question of how much you *want* to know," the priest said. "Percentage-wise, if we were to use that term, you have seen more death in your family than your average hundred-year-old. When your parents passed away, I couldn't understand how God could be so cruel. It was straight out of the Old Testament. An act of vengeance. Followed by those strange days when you were missing. Sverre stopped the tractor in the middle of the field and went straight to France. Took the first flight available, which cost a small fortune. I prayed for you six times a day. Only God knew where you were, and I wonder still if He is the only one who knows the truth. Then you were found. And I saw God's light by the produce counter at the general store. It shone on a young boy and his grandfather. I am telling you this to console you, Edvard. The truth is that *you* were Sverre Hirifjell's deliverer."

He said it was consolation, but I began to dislike the direction it was taking. It was as if he were speaking to the leader of the parish council about someone else. The combination of loose tongues and kindness which has always given Christians an excuse to dig about in other people's affairs.

The priest began to talk about the post-war period. About how my father could not stand being here on the farm, because he blamed *his* father both for his Christian name and for the resentment he inherited. "Walter got beaten up at school," the priest said, "because his father had been on the wrong side. He travelled to Oslo and got a job when he was fifteen. During the entire post-war period, Sverre and Alma muddled about here, kept to themselves. Never went into town. They were stared at and slandered in Saksum, even by churchgoers."

I realised all of a sudden why the old kitchen garden at Hirifjell had been so big. Why everything was organised to be self-sustaining, with a henhouse and pigsty, a rabbit hutch and stalls for cows, sheep and goats. It was because Alma hated going to the shop. The reason Bestefar always bought expensive things, preferably German, so that he could avoid going into town for repairs. For many years they didn't even subscribe to the newspaper, Thallaug told me.

"I know you have always believed that the village held a grudge against Sverre," the priest said. "But that hasn't always been the case. Everything changed when he adopted you. For the first time in twenty-five years he began to show his face in the town centre. A stubborn old man taking on a three-year-old. No matter what Sverre Hirifjell might have done during the war years, people's perception of him changed when they saw the two of you together. People realised that Sverre Hirifjell had never done anything directly bad. He never informed on anyone in the resistance. He had served on the Eastern Front, and his choices of car and tractor were German, but nothing more than that. There was only the occasional incident afterwards. A bit of foul language when bringing the sheep down from pasture, trivial matters, like that fracas when Jan Børgum painted the swastika on his car door."

He followed my movements. Studied me as I cut the potatoes, as I reached for the salt shaker. I had a mouthful of peas on my fork, but my hand stopped halfway and our eyes met.

I did not agree with his reference to "the occasional incident". It had shaped my life more than the German Mauser hidden away in the loft. For as long as I could remember I had been standing up for my grandfather. It got serious at lower secondary school. The history lesson when Halvorsen said what he said about Bestefar. Rather, he did not say it about him, but the whole class knew that every word applied to my grandfather.

"The Front Fighters," Halvorsen said, "they may not have known better. But they betrayed the true government and served the Germans."

Halvorsen commuted from the neighbouring village and was my form master from Year Seven. From day one he was so bloody obdurate. Even if there was an opportunity to slip in a "maybe", he never did. In history, all he ever talked about was war, and *the war* in particular. Stood there in his grey overall coat, with his disgusting eczema, and when he could have said Front Fighter instead of traitor, he would throw in fifth columnist or collaborator to boot, and then gloating he would add that *quisling* was a word adopted by other countries after the war.

He went on like that, his coat pockets white from the chalk on his hands, the bloody chalk he used to write *The Truth: Terboven, N.S., the liberation, the purge of traitors,* words which were dotted with spittle when he harped on at his worst.

According to local gossip, Halvorsen's father had been tortured. Even so. As a teacher he could have delicately slipped in something about people being young, that it had not been easy to choose. That there were plenty of people who, when the country became safe in 1945, were suddenly brave enough to run around with clippers, shaming girls who had made mistakes too by shaving their heads.

But Norwegian History was as it was. Saksum lower secondary school 7A could not jump from 1940 to 1945 just because I was in the classroom.

I remember the day it came to a head. I was sitting in the row by the window, the spring sun shining, bare asphalt, the ice about to break on Laugen. Halvorsen rattling on. Gave me a sidelong glance, as if to see whether I was about to crack.

"The Norwegian Legion. Can anyone explain what that was?"

I was aware that several people had raised their hands. The clever girls by the door. A few at the very back. But Halvorsen said:

"Edvard. Are you with us?"

Us.

"The Norwegian Legion, Edvard. What do you know about that? It was in the passage."

"What do I know about the Norwegian Legion?" I said.

"Yes, that's what we are asking."

"We?" I said.

"What do you mean by that?" he said.

"You're asking as though everyone in the class is on your side," I said.

"Regardless, Edvard. What do you know about the Norwegian Legion?"

"I know more about it than you do."

"You had better give me an answer, Edvard. What's your opinion on the subject?"

"That you should have been there yourself, you bloody know-all."

Then I dashed out of my chair, trying to keep myself from crying, but I was sobbing by the time I reached the door, and as I ran down the tiled corridor I no longer gave a damn that everyone could hear me.

But this was not something to whinge to the priest about now. Instead I said something else entirely, without managing to ask myself whether it was appropriate, it slipped around all barriers like a mutt that wants to get past the fence and is just waiting for the opportunity.

"Did you meet with my mother often?" I said.

He was not taken aback by the question. Did not crack his fingers or scratch his jaw, said simply, "A few times. When I heard that Walter had met a French girl and moved back to the farm, I came here to introduce myself. Nicole, yes. She didn't say much. Was shy, as I remember. Spent a long time in the cowshed even though she knew the priest was on the farm. But when finally she arrived, well, I won't forget her face. She was always looking around her, as though everything was fleeting. A deer on alert. You're like her. The mouth. The same eyebrows. You have her hair, too."

"I don't even know how they met," I said.

"She came here – as a kind of tourist."

"To Oslo?"

"No, I believe they met here. I know that Sverre held Nicole dear. Alma was more guarded, you know. Nicole was – no, we should concentrate on the funeral."

"She was what?" I said. "Tell me."

He cleared his throat and shifted in his chair. "No, it was nothing really. The first time I met her she could only speak a few words of Norwegian."

"And what about later?"

"Hm?"

"You said the first time. What about the next time?"

The priest began to pick at a chip in the coffee cup. I saw his face change. He straightened up and stared fixedly at the table as if he were looking at a bible, one he would consult before a sermon even though he knew exactly what was written inside.

He said nothing more.

I wanted to hear about Mamma. But it was a painful and shameful thing to do. To ask someone else what your mother was really like.

"The funeral," the priest said. "Have you heard about the coffin?"

"Rannveig's son let the cat out of the bag," I said. "Had Bestefar already chosen a coffin?"

"It has been ready for many years."

"He never said anything about it."

"That's because he didn't know."

I stuck the knife through the potato so that the steel clattered against the porcelain.

"He didn't *know*?" I said.

Thallaug shook his head.

"So who arranged it? Alma?"

The priest rubbed the corner of his eye. "It was Einar. He made a coffin for his brother."

"In case Bestefar died on the Eastern Front?"

"No, it must have been later."

Something slipped, he began to wipe his glasses with the same handkerchief he had used to blow his nose. I worried then that he was going senile, that he was going to bungle the funeral.

"Tell me," the priest said. "You take photographs, don't you?"

"Yes." I couldn't bring myself to ask how he knew. Could Bestefar have talked about me?

"There's a bit of Einar in you," the priest said. "He could capture the form of something he had seen and use it in another context. Einar was completely different from Sverre in this respect. Einar interpreted everything he experienced, he was a thinker and a dreamer."

"But when did he make the coffin?" I said.

His gaze grew distant. When he answered, it was as though he had not grasped what I had said.

"Einar, he disappeared from us. Twice he disappeared. The village's foremost cabinetmaker. One of the best in all of Gudbrandsdalen."

"Including Skjåk?" I said.

"Including Skjåk."

"He disappeared *twice*?"

"Hm," Thallaug said. "This will take a while. What time is it?" He fished out his glasses case and put on yet another pair of spectacles.

"Almost three," I said.

"I have to be back to take my pills at four."

"I'll remind you."

"Please do. Otherwise my ministry will expire at a quarter past."

The priest began to tell me about the farm's prodigal son, and when he talked about Einar, it seemed as if he was also talking about me. Not precisely, but about the boy I had often dreamed of being. Except that the sketching pencil was replaced by a camera, the cabinetmaker's workshop by a darkroom.

Einar could not be bothered to pay attention in lessons. In a notebook from his confirmation studies, his sentences would stop in the middle, but in the margins there were sketches of furniture, houses, cities, more furniture. "What could I say?" the priest said. "That he had to stop? A young man in Saksum in 1928 dreaming of craftsmanship? Luckily his parents accepted his talent, even though he held the allodial rights, and they sent him to the academy at Hjerleid to learn the trade of cabinetmaking. His talents were staggering, and his desire for experimentation was so exceptional that even his most ambitious teachers realised that they were holding him back."

The priest went on to tell me that, after a couple of years, Einar tired of acanthus leaf patterns and the country house style. So he went to Oslo to serve as an apprentice, but grew bored there just as quickly. By then he had long since begun to sign his pieces with a small squirrel hiding its nose in its tail, a mark he used for the rest of his life. In 1931, at barely seventeen, he raced off to France to find work.

"We heard nothing more from him," the priest said, "but later I learned that he had been at Ruhlmann's for many years, one of the foremost furniture designers in Paris. Einar was one of his master cabinetmakers."

"Did he stay in France for a long time then? I thought he was there only briefly?"

"No, Einar came to be considered a Frenchman. They did not use this term in those days, but he employed a style that was later called *art deco*. I thought, alright, Einar has found his place in this world. Meanwhile the farmer in Sverre grew, he and Alma were married and ran Hirifjell without the allodial right being clarified. Listen, Edvard: did Sverre tell you much about this time?"

"Practically nothing," I said. "For us, time began when I was four years old."

The old priest drank his coffee in the old-fashioned way. Allowed a lump of sugar to break the oily surface, waited until it turned brown, placed it between his lips and sucked it before pouring coffee on the saucer and taking a slurp. I noticed something similar happening inside him. The priest was separating things, so that only what tasted of sugar emerged.

"Just before Christmas 1939," he said, "Einar appeared at my door and said that he intended to move back to Hirifjell."

I stopped chewing. This was news to me. Einar must have stood right *there*, in the doorway. I pictured him carrying a suitcase and a select collection of tools, and how the couple must have gaped when their supper was interrupted.

"Einar's homecoming did not exactly please Sverre and Alma," the priest said. "For one thing, the first-born son had returned, with no knowledge of farming, and hardly suited for it either. Einar had been bad at replying to letters and had not come home for the funeral of his father, the man who had actually paid for his training. Then there were his habits. Here was a boy who, already at the age of fourteen, had thought that this was a narrow-minded valley with no place for big ideas. Imagine it – nearly eight years in Paris in the thirties, then add a measure of dangerous cockiness on top. Not that he looked down on village life, but he wore a wristwatch that could be flipped so the glass faced down. People in Saksum did not realise that was what it was, they thought he was wearing a bracelet. His hair was styled into a kind of strange swirl with little curls across his forehead. And then there was Alma and Sverre, each with their own N.S. membership, slogging away for fourteen hours a day. Still, Einar made no great demands, he simply asked for a parcel of woodland to source materials from, and to be able to extend the cabinet-maker's workshop."

"He got that," I said. "Above the fields there's a small plantation of birch."

"They are beautiful, are they not? He began cultivating trees like that when he was only thirteen. The office furniture at the parsonage is made from those materials. I ordered them in 1939. As a believer, I am hesitant to use the word 'miracle', but the radiance of those tabletops draws it out of me. And as far as profound sights go, there is little that can measure up to the deep

47

patterns in the wood. It's like staring into a fire. You always see a new face in the flames. I told Einar that when I received the desk. In response, he made me a chessboard, which I have to this day. Flame-birch for the white squares and walnut for the black ones. That is also how I view you, Edvard. Your father and your mother. The south and the north. Darkness and light. Your internal struggle."

"What do you mean by my 'internal struggle'?"

"It is clear from a distance but vanishes in a mirror."

All of these intimations. How deep inside me could he see? The only close contact with him was before my confirmation, and that was in the years when the loss inside me had been tearing me apart. At the time, I would hand in blank exam papers and cut class. I took the bus to Vinstra and left my rucksack at the bus stop. Hitch-hiked to Rockestugu in Otta and bought all the records I could afford. Or, if most of the cars were heading south, I would cross the E6 and catch a lift to the snack bar in Skurva, before trudging down to the business park and asking for car brochures, saying they were for my father. Went to Stavseth, Skansaar, Motorcentralen, all of the big dealerships. Invented a father who drove a Citroën D Special or a Ford Granada, depending on how I felt. Or went to Melby's sports shop in Ringebu in the middle of the day and looked at air rifles, picked out fishing rods and said that my mother had promised me five hundred kroner for my birthday, and that I was just looking.

All of this the priest must have seen back then, but that was water under the bridge now. I had struggled through. But *clear from a distance*?

"What were they like?" I said. "Einar and Sverre. When they lived here together."

"They were different from the day they were born," the priest said, and poured himself more coffee. "But it only came to a head in the spring of 1940."

"When the Germans arrived?"

He nodded slowly. "A long, dark column. Ugly, angular

vehicles. They were sent to head off the British up near Kvam, but they were worried that the motorway had been laid with mines. So they drove along the upper village road at a frightening speed. They passed right by the church. I was sitting in the sacristy when the walls started shaking."

The priest waved his arms around and described how he had heard a tremendous crash from the choir room and that three hundred hymn books had fallen off the shelves. He thought that now he was going to rise up and meet his Maker. But out in the nave, he saw that it was the altarpiece and the great crucifix which had come loose and crashed to the ground. They had hung there for hundreds of years, survived the great Storofsen flood and the forest fire of 1748. Now the altarpiece was broken and the cross was snapped in two. The figure of Jesus was split at the navel, the face was crushed and the arm hung limply. He heard another column approaching and ran outside.

"There they came," the priest said. "Grey trucks with caterpillar tracks and iron crosses on the sides. The church shook again, the chandelier rattled and Jesus lay there with a broken back. I took Our Saviour into my arms and raced outside. The soldiers were jittery and their rifle bolts were rattling when a flatbed full of infantrymen aimed their Mausers at me. I held out a broken Jesus and shouted in German that they had best slow down if they held out any hope that the words on their belt buckles – *Gott mit uns* – would still apply."

"But surely they were not bothered by that?" I said.

"Oh yes. They were terrified, that goes without saying. They were only a few hours from the front. I knew already as a young priest that a cross has the power both to comfort and to frighten. So they slowed down and had soldiers guide the traffic more slowly past the church. But myself, I had to hurry. I believed that with the outbreak of war, even my parishioners were likely to turn to God. If people saw that the altarpiece of the church in which they were christened was crushed, they would lose all hope. I carried the crucifix back inside and locked up, then cycled here to Hirifjell in my vestments to find Einar. He and Sverre were sitting in the kitchen. Even then I could see the hostility between the brothers. They were talking in raised voices. Sverre was certain that the Germans had come to defend us against an English invasion. Einar said that he was going to follow whatever the King and the government recommended. I stood in the doorway holding a Jesus in need of wood glue. Einar followed me outside and I explained the situation to him. He filled a rucksack with tools and all the screw clamps he had in his workshop, and worked through the night in the church. The wood was brittle and many of the fragments could not be found. He cut tiny little pieces, coated them with glue and matched the colours. At an extraordinary speed he produced slivers of wood the size of spruce needles. He put Jesus back together and gave him his face back. Well. The Germans arrived on a Saturday morning. When we rang the bells for Sunday service, the altarpiece and the crucifix were in place. Einar slept in the sacristy while I conducted what I later realised would be my best-ever service. The only one not seen in church was Sverre."

I finished eating and pushed aside the empty plate. Felt more alone than ever. This was something *I* should have retold, this should have been *my* family story.

"When did Einar go to Shetland?" I said. "What year?"

"1942."

"Sverre was still on the Eastern Front at that time?"

"Yes. Einar left the farm only a few days before Sverre came home. The plan was to join the resistance. Strange, I thought. Because Einar was no fighter. Nor an idealist. In debates he would stand in the shadow of his brother, a man who went to meetings and demanded action."

"Did you hear from him again over the course of the war?" I said.

"Not a word. In 1944 I discovered that the rights to Hirifjell were to be transferred to Sverre. Then came a report from the Germans, that Einar had been shot in France. He was said to have joined the resistance there but was executed by his own people. I still have the letter, with the Imperial Eagle and all. The date of his death is written in the church register, the line where the ink is smudged. Yes, I cried. But in 1971 I opened the book once again to enter the death of your father and mother. That was when I thought back and began to dig through the archives. Two things have troubled me since that time. The first is that Einar Hirifjell was shot in Authuille, the place where you disappeared and where your parents died."

"What are you *saying*?"

He wrinkled his brow and scratched his ear.

"The second thing was, how he could have made a coffin after his death? Einar might have been shot in 1944. But the bullet could not have struck his heart or his brain. Because in 1979 a lorry parked outside the funeral parlour and delivered a magnificent coffin. It was made of flame-birch and had been sent from the Shetland Islands. Go down to the funeral parlour and you'll see."

3

A WHITE MANTA TURNED OFF THE COUNTY ROAD. AN
entire summer with no visitors, and now that he was dead the
cars were pouring in. Did they come because he was dead, or
because he was dead?

The headlamps shone through the grey weather and made the
wet clumps of grass along the verges glimmer. The rain was falling
once again. Actually no, it could not be the same rain. Thoughts
like this had crept into my mind ever since the priest left, and I
paced around the log house, my head reeling.

I leaned against the door frame and wondered what he wanted
here. It was only when the Manta got closer and the wipers swept
across the windscreen that I could see that she was driving her
brother's car. He was now doing his national service in the north.
The headlights dimmed, she opened the car door and sat there
with the music spilling out. Cowboy Junkies. "Blue Moon". I knew
that trick. Her way of expressing her mood without saying a word.

She was even more attractive, wearing a light-yellow dress.
That was not like her. When she lived in Saksum she rarely
dressed up. Just threadbare Levi's that accentuated her nice, slender bottom. No hair dye, never any make-up. Sensibly dressed,
usually from the spring collection when it went on sale. But she
had firm thighs from handball practice and the hollow of her
neck would glisten in the summer sun, and she could be tough as
nails when necessary.

"Just come inside," I said. "Don't sit there showing me how
easy it would be for you to leave again."

She strolled confidently into the living room, stood and looked at the framed photographs above the couch.

"Is that from this year?" she said, pointing at the picture of street lamps in Saksum at night. I had strapped on my skis one evening and herringboned up to a steep overhang on the hillside, a tripod and camera in my rucksack. I set up the Leica, sat there until it got dark, and when a lone vehicle appeared I set the exposure time for thirty seconds. The town centre was bathed in a yellowish light and the rear lamps of the car became one long, red streak heading south.

"I took that two years ago," I said. "You can see the new building at the school isn't built yet."

The television was on. She switched it off and walked out to the conservatory.

"Who told you?" I said.

"Someone sitting next to Garverhaugen at the café. He'd been out fishing for grayling and had seen the police chief, then the doctor, and then Rannveig Landstad driving down the road on the far side."

"But he didn't see the old priest?"

"He must have already settled himself in the café by then. Tell me, Edvard, how are you doing, really?"

"Surviving, I guess. The worst part was that bloody mess with the swastika."

"I thought you had grown out of all that," she said.

"I wouldn't throw a punch someone said 'Nazi' for the third time."

We had been through this a number of times. At village parties I would defend Bestefar, and that would end in a row with the two of us yelling at each other, arguments in the fireweed at the side of the motorway in the middle of the night, after we were forced to walk home because I could not stand the people who had offered us a lift.

We went our own ways that spring when she finished upper secondary. I went into the potato fields, she went to Oslo. Made new friends. Did well in her exams.

"What are you going to do now?" she said.

"He's not even in the ground yet, Hanne. Don't start. Not now, at any rate."

I knew that if she did not stop, the old argument would start up again. She would repeat the same thing she had gone on about on so many occasions. That I would never go anywhere. Would never *get* anywhere. But what about her? Oh yes, she changed when she went to Oslo, but that consisted largely in buying boots with metal caps on the toes and heels and wearing tight jumpers under a leather jacket. Inside was a girl who had always set her sights on returning to the village. Her education took her only far enough to reward her with a steady job back home.

And who was I to criticise that? Was I supposed to go around moping, demanding everything of others and nothing of myself? She was right to ask me the same question I had asked myself earlier that afternoon when I put the coffee on and made a full pot by force of habit. *What do I do now?*

Well. All I had to do was look out of the window.

Go over the potato fields with a ridging plough. Change the diesel filter on the new Deutz. Jack up the sagging wall of the outbuilding. Drive into the mountains with salt lick for the sheep. Change the gutters on the south side of the barn. Clear the weeds from the vegetables. Find out why the rotovator wasn't starting in the heat. Arrange Bestefar's funeral and spray the fields to keep the blight at bay. I should have done all this during the week, because next week was the only chance I would have to change the windows up at the mountain pasture, if the weather held up.

Only one thing stood out: I had to pay a visit to Rannveig Landstad and see the flame-birch coffin.

Hanne came to me and held my head in her hands.

"Poor you," she said. "It is not even showing on you."

"I feel it, though," I said. "Deep in my gut."

"But you look just the same as before. Maybe you've been through so much suffering that you don't have room for more."

That was all it took. I pulled away, sat with my head against the wall and wept. It surged upwards like a flooded cellar. How would things have turned out if I had had someone? What kind of a boy would I have become with parents, brothers and sisters even, with young people around me, relatives who thought I was worth spending time on?

My hands and feet ceased to be a part of my body. I felt like one gigantic heart, a swollen shapeless lump pumping out tears that had waited for twenty years.

I sobbed and sniffled for an hour. When I had finished, I was as exhausted as if I had walked from the village up to the mountain pasture.

She stood looking at me. No accusations. No fake sympathy.

Just the question she had no doubt been fumbling for when she was at veterinary college. Whether a boy and a girl got together because they really *suited* each other, or whether in small biotopes like Saksum they chose whoever was available and simply got on with it.

"Hanne," I said. "I have something for you."

I opened the glass cabinet and took out the pearl earring.

"Gosh," she said, holding out her hand. "*That*. After such a long time."

She stepped closer to me, and her city ways and her distance were gone. I recognised the girl who had been with me at Hirifjell that summer, and I remembered one exception to her never dressing up. She had done it then. That summer when Bestefar took what he called his "office week", and went to the annual meeting of the Norwegian Association of Sheep and Goat Breeders.

Ever since I was thirteen, Bestefar had gone away for a week each year and I had to "run the farm on my own". For me it was an adventure. I cycled to Laugen and fished for grayling, prepared my food, kept my promise not to drive the new Deutz, or play with matches. I only had to be sure to be home between five and six, when he would call to make sure the farm was still standing.

During that summer with Hanne I wished that the annual breeders' meeting could last for three weeks. She was fifteen and had done as she wished for the past year. I remember every single day, how they were bursting with *us*. She dressed up then. Alone on the farm. We woke with Flimre, the striped forest cat, lying between us. Without exchanging a word we would pretend that we had a child together – the size, the weight, the warmth.

We were grown-ups when we felt like it and children when it suited us. We adopted ways of conversing, drank coffee in the morning and beer from the cellar in the evening, we bought proper cigarettes and took three drags each. Neither of us much liked smoking, we only did it because we saw it in the movies. And because it felt right to smoke a Pall Mall after sex. Rollies would have cheapened it.

I remembered her, pure and beautiful, wrapped in a sheet in front of the window on the first floor, and I knew that she was taking in the view of Hirifjell, the vast sight to which only the eyes of a young girl or a Leica could do justice: redcurrant bushes dense with berries, the flag-stoned path leading to the swimming hole at the river, the creek which cut through the potato fields and disappeared from sight behind the barn. The fruit trees, the pea pods that dangled like half moons when we got close to them, so plentiful that we could fill up on them without taking a step. The dark-blue fruit of the plum trees, the sagging raspberry bushes just waiting for us to quickly fill two small plates and fetch some caster sugar and cream. The old tractor and the new tractor next to each other, their wheels freshly hosed down.

I saw it then, how earnestly she viewed it. At Hirifjell there was no mess and none of the half-finished chores the farmers in the village waded through, year after year, until they were blind to the tractor tracks by the front door, the rusted hay forks left untreated for the tenth year in a row, the cracked slurry tanks that could be seen from the motorway. Hirifjell was a model farm with edged grass right up to the white foundation walls and a swing that swayed in the wind.

A farm she could grow into.

It was well into the third day before Hanne noticed that her earring was missing, no doubt because she was so unused to dressing up. Now she took it and rolled it firmly between her fingers.

"You've had it the whole time," she said. "You have."

I shook my head.

"It was in his chest of drawers. In Alma's old jewellery box. He must have found it and thought it was hers."

"Edvard. You haven't started sorting through his things, have you? Already?"

"I need to have something to do."

"Sorry to be blunt, but that was quick."

"Put the earring on," I said.

She stepped back. Placed one foot against the wall so that her bare knee was pointing at me.

"Don't expect anything," she said. Tilted her head, used both hands to fasten the earring.

"Hanne," I said afterwards as we shared a Pall Mall. "You remember Einar? Sverre's brother?"

She sat up in bed, held the cigarette pointing upwards so that the ash would not scatter. Blew a lock of hair from her eye.

"The one who built the workshop?" she said.

"Yes. I think he's still alive."

"That can't be."

I told her about the coffin.

"How old *is* the old priest?" she said.

"Almost ninety."

"There you have it."

"No. He's clear-headed. More or less. But he knows something about my mother that he won't tell me. About Einar too."

Hanne handed me the cigarette and rolled out of bed. Got dressed with her back to me. It was hopeless to mention my family history to her. She was a girl for the good things in life. A girl for Easter sun and red skiing gaiters. For shiny Bunad brooches on Norwegian National Day in May.

We stood outside. Hanne ran a finger through the raindrops on the boot of my Commodore. I looked up at the log house, the yellow light from the living-room window falling onto the black-currant bushes. A solitary light on the second floor. He must have forgotten to turn it off when he was up there.

"You're right," she said. "Let's start."

"Start what?"

"Taking off his sheets."

"Right now?"

"Well *you're* not going to be able to do it. Let's get it cleared out."

The log house already smelled dry and stuffy. The fridge was open and unplugged. It was the only sensible thing I had managed to do after the priest had left, taken the food and transferred it to my fridge, even though I could just as well have left it in his and fetched things as I needed them.

In the living room the newspaper was still at the foot of the couch. There was dry sand from his shoes on the paper.

She went upstairs. I heard the click of the turn switch, the creak of floorboards. Her steps. So much lighter than Bestefar's. She came downstairs with the sheets gathered in a huge bundle, leaning against the banister because she could not see the steps. "I brought down a load of washing too," she said. "Is the washing machine still in the cellar?"

I began to think about her in a new way. About life with a girl who felt her way forwards with her hips.

"Let's throw them out," I said. "No-one's going to use them anymore."

"Use them? They were your grandfather's."

"They are the bedsheets of a dead man."

She rubbed the material between her fingers. "This is fine linen," she said. "If you don't want them, I'll take them."

"You can't mean that."

"Sverre was always nice to me, even though he knew what we were up to. One time I came up here and you weren't home, so he offered me a crushed nougat ice cream. Said that it was nice to see womenfolk at the farm. Even though I was only fourteen and driving a moped illegally."

"I don't understand," I said. "You haven't been here for three years. And now you're making yourself at home."

She shrugged.

"You're here because you feel sorry for me," I said.

"And?"

"Stop it," I said and carefully picked up the newspaper so that the sand trickled towards the fold. I pushed open the front door with my shoulder and tossed the grains outside, as if they were ashes from a cremation, as if the door frame was the gunwale of a boat, as if the farmyard was the Atlantic Ocean.

She brought fresh air with her. Undid the hasps on the bedroom windows. Opened doors, created a breeze, let in the fragrance of summer rain. It was not the work I took note of, but the way she filled the house. Her steadiness, which before had seemed staunch and old-womanish, now gave way to something freer, like a view that appears when tall trees are felled.

But when she opened the sliding door of the wardrobe in Bestefar's bedroom, the one that covered the whole of one wall, it grew cloudy inside me again. The darkness of the wardrobe suggested something dusty and dim and old. Clothes without a body.

A sudden sensation emerged from within me. A memory, I could not tell whether it was real or not. Of my mother dressed in something blue.

Hanne reached into the woolly shadows of the wardrobe and pulled out the smell of being old. Piled a bundle of clothes onto the bare mattress. Faded shirts, string vests, work clothes. Grabbed another armful. Wrinkled her nose, leaned further inside and unhooked a black suit cover with a zip.

"What in the world . . ." she said.

Even I could see that it was an expensive suit. Tightly woven fabric without a wrinkle. Light-grey pinstripes on dark-grey worsted. A cut that could make a man look like he owned a bank. She looked at the label on the inside pocket and pointed at the name of the tailor: ANDREAS SCHIFFER, ESSEN.

"Edvard," she said. "Could this have been your—"

"No," I said quickly. "My father was taller. And all skin and bone."

"This is an expensive suit," Hanne said. "I mean, *really* expensive."

She took the jacket off the hanger and held it up to me. I stepped back and shook my head.

"Are you sure it wasn't meant for you? As a present?"

"Neither of us had any interest in clothes. You've said as much yourself."

She checked the pockets. The lining shone faintly in the lamp-light as she pulled out a light-blue ticket. I leaned over and we read at the same time:

Bayreuther Festspiele.
Vierte Nacht: Götterdämmerung.
Samstag, 30. Juli 1983

A shudder passed through me, through both of us. She recognised the date, it had meant something to her too. The summer we were alone while Bestefar was at his annual meeting.

Or not. I had never wondered why the Association of Sheep and Goat Breeders had to have such lengthy annual meetings. When he telephoned me I might have wondered why the line was so bad, but back then I probably thought that the meeting was so far away that a little rustling and crackling on the line was nothing out of the ordinary.

Götterdämmerung. I remembered when he had picked up the enormous box of twenty-two records at the post office. It had cost thousands of kroner. Carefully, with the sharp edge of his bayonet, he had slit open the wrapping paper from Norsk Musik-forlag, put the box on the living-room table and said:

"You see, Edvard. '*Der Ring des Nibelungen*' is the only piece of music which can stand alone."

Now I grabbed the suit, as if tearing the jacket from a thief, and went through the pockets to see if there was anything else.

There were more tickets in the other side pocket. "*St John Passion*" in Hanover. "*Tannhäuser*" in Munich. "*Missa Solemnis*" conducted by Karajan, five movements by Bach on the Hilde-brandt organ at Sangerhausen. The dates coincided with an empty chair at the annual meetings of the Association of Sheep and Goat Breeders.

Amongst the concert tickets was a crumpled receipt written in copperplate on thin paper. The letters were smudged with moisture and the only thing legible was HOTEL KVELDSRO. It sounded like a guest house somewhere near the coast.

"Maybe he went to a funeral," she said, as if to gloss over the quiet deceit we had discovered.

"Someone he shared a trench with on the Eastern Front, you mean?"

She rubbed her eye. "Does that matter?"

"Why didn't he just tell me?" I said. "To my face. That he felt like seeing a Karajan concert, and that he'd be away for a week?"

"Maybe he wanted you to feel that the farm was yours," she said. "Give us some time alone."

"Or he just wanted to hear '*Tannhäuser*' in peace," I said.

"What do you mean?"

"It's so peculiar. Everything we did, we did together. But that was just farm work. Never any trips. It's as if he was worried I might find something to take me away from him."

"Is there anything that could have taken you away from him?" said Hanne.

Was she *pretending* to be blind? If that was the case, she was about to take on the role of the one who had bound me to Hirifjell.

"Sverre could have been away for three weeks as far as I was concerned," she said, stroking my arm.

"Well, now he's in the concert hall for good," I said.

I stood looking at his suit in my hands. As if I was holding his shell. All at once I remembered his footsteps. "He was doing something upstairs last night," I said and put the suit down.

Then we were there, in a drawing room that was alien to me. The hallway on the second floor had been dark as a mine shaft for as long as I could remember. The curtains drawn, the light bulbs dead. But now a yellow ceiling light shone down on the barren room, and on an open secretary desk in the far corner.

"All those papers," said Hanne. "He must have been looking for something."

She leafed aimlessly through the sea of envelopes and documents. Receipts for tractor equipment, old tax returns. "There are some slides here," she said and handed me an orange plastic Agfachrome box.

"They're just empty boxes," I said. "He always transferred the pictures onto glass plates. They're down by the projector."

She held up a transparency to the light. "There are pictures in this one, at any rate."

I was taken aback. Bestefar had no particular interest in photography, even though he helped me through all 230 pages of *Leica Technik*. He was content with a Rollei and used exactly one roll of film a year, always twenty-four exposures. But from each box Hanne now picked out twelve slides in cardboard frames. I took my pocketknife, broke loose the film itself and looked at the numbering.

Bestefar had used one film a year sure enough. But thirty-six exposures, not twenty-four. The other twelve were taken during his secret week abroad.

So that was why we never looked at the year's slides straight away. When the Agfa package arrived from Sweden at the end of the summer, he always told me I had to wait, and then he would go up to the box room and slot the pictures into glass frames. Then we would draw the curtains, switch on the projector, and in the dusty cone of light between us and the screen, we would view the year we had just spent together.

Hanne handed me one transparency at a time. Bestefar's extra pictures tallied with the concert tickets. Freshly swept pavements and half-timbered town halls. Opera houses, the coulisses in Bayreuth.

I imagined him in the one week of the year when he could stroll about and either understand or feel understood in Germany, a proud man in his sixties in a charcoal-grey suit from Andreas Schiffer, alongside all the others who had lost the war.

We started on the rest of the boxes. All the photographs seemed to be of Germany, and one stood out. It was numbered 18b and was so different that it could have been taken by a different photographer. One year, impossible to tell which, Bestefar had taken a single photograph of a bleak, insignificant coastline with a small island on the horizon.

"Edvard?" Hanne said quietly. "Look at this."

I got up from the floor and reached for the five envelopes she was holding. Written in Bestefar's calligraphy. *Walter. Nicole. Alma. Einar. Edvard.* They were sealed, all except for mine. Like mystery bags for the dead.

"Should we open them?" said Hanne.

It felt as if I was holding five rounds of live ammunition. Mamma's envelope was thin, Pappa's even thinner. Something slid around in Alma's envelope, a small book, perhaps.

"You're sweating," Hanne said. "Are you alright?"

I was aware of her touch, but all I could think of were the five names. Bestefar had a long time ago prepared these envelopes, and then waited until I was old enough.

Or until he was.

"Let's go downstairs," I said, and put the envelopes back.

She turned in the doorway as we left the room. It was as if she were looking for an excuse to stay longer. Down on the first floor she went back into Bestefar's bedroom.

"What are you doing?"

"I have a hunch," she said, and began to rummage through the wardrobe. I heard a cardboard box being moved, the rustle of tissue paper.

"Look at this!" She held up the wedding dress. "Look at the lacework. What craftsmanship. My goodness!"

"Must have been Alma's," I said.

She pinched one sleeve and stretched out the fabric, arched backwards and held the dress against her chest, stared down at herself, at the way the material curved over her breast.

"Close your eyes," she said.

I was about to say no, but sat down on Bestefar's bed with my eyes closed all the same. It was as if I had set out on a journey, the course set in stone, and I could sense something knocking in the far reaches of my memory. Something was not right, something to do with the two of us.

I heard her clothes fall to the ground, cotton on skin and the swish of silk, I heard her hold her breath and then exhale, before the whisper of finely woven material moved through the room.

"Look at me, Edvard."

She stood over me as if about to straddle me, her face hidden behind a fine-meshed veil, her skin taut at the collarbone, white tulle against her breasts, her hair curling over her cheeks.

I held back my confusion, disguised it as arousal.

She straightened up, and I felt a tightening in my stomach because I knew she would stand like that one day, maybe soon. She was going to walk down the aisle of Saksum church, and the man waiting for her could be me. And from then on I would be the perpetual potato farmer, Edvard Hirifjell.

"Put on the suit," she whispered.

We stood next to each other, me in an Andreas Schiffer suit, and she so absorbed with what she saw in the mirror that she would have remained like that even if the house had burned down.

"Imagine," she said. "We could have been *them*."

"No. I can't cope with that."

"Yes, you can. This is you. As you could be."

I looked at the two of us in the mirror, at how she devoured this moment like a marzipan cake. My own eyes eating up my reflection.

It smelled of evening. I followed her tail lights from the upstairs window. Watched the twilight covering the farm buildings and fields. I had been the boyfriend of the caterpillar, and now the butterfly was soaring.

I walked over to the secretary desk and fanned out the envelopes.

Nicole. Resealed with yellow tape.

Walter. Same tape.

Alma. Resealed with fresh tape.

Einar. Resealed with freezer tape.

I took a look inside my own. Vaccination card. Report from primary school. Report of criminal damage after a fight at Venaheim community centre, where I wrecked a door. My birth certificate. The name impressed in typed letters, Edvard Daireaux Hirifjell. Could that be right? Hirifjell was the only name on my tax return, and whenever else the authorities needed to contact me.

The signature I recognised. Our entire family history had passed through the fountain pen of the old priest.

I began flipping through the documents for the farm. I wanted to find something Bestefar had written, something stating that Bestefar was Bestefar, and no more. A dependable man who kept his faith in orderly personal records and impeccable correspondence on an Adler typewriter.

Tractor/equipment 72–75. User manual for a harvester we took to the scrapyard last year. The carbon copy of a complaint letter to Fron Traktorservice for the old Deutz. It started locking in reverse the week after the warranty expired.

I was among the first to buy a Deutz from you, and have since been loyal to the brand, something I will continue to be if this tedious matter with the gears can be taken care of amicably.

The potato harvester, every single litre of agricultural diesel, tractor equipment bought at Ottamartnan. Receipts for the sale of seed potatoes to Strand Brenneri. The desk was brimming. Were the hundreds of kilos of old documents intended as a barrier to my curiosity? Until he gave in and unlocked the desk last night?

I cut open Alma's envelope. A long history of illness told through letters. The results of an X-ray. A copy of a letter she had sent to the district physician.

There was the bill for her funeral. Coffee and cakes for fifteen at the guest house in Saksum.

A large almanac which she had used for almost ten years. She had begun it in April 1961, and the last entry was from 1969. I flipped through the years. It was primarily a journal for the farm, for sowing and harvesting, lambing and slaughtering. Some figures I did not at first understand, until I realised that it was her weight, month by month.

I remembered her skin, the rough cotton of the apron she had made. Once she had been a woman who wanted to keep slim.

On the final pages she had written birthdays and telephone numbers. Some names were crossed out with *Deceased* added in a different colour ink.

When I got to the notes for 1967, I found one line which stood out. It was written crosswise near the margin, so close that the rusty staple had stained the ink.

Einar. Lerwick 118.

In 1967? There was no annotation of *Deceased.* Despite the fact that our entire family history had confirmed, time and again, that Einar had died during the war.

Could 118 have been Einar's post office box?

I put the almanac aside, opened a drawer and took out some frayed bundles of paper. They were held together with twine, the years indicated in pencil on sheets of white paper wrapped around them. The same system from 1942 until now. Bestefar's life did not fit into a notebook.

I sat on the cold drawing-room floor, flipping back through life at Hirifjell. The years grew ever darker. Rejection of compensation for fire damage at the mountain pasture. Verdict of treason in 1946. I cut open the war years. Membership cards for Nasjonal Samling. Envelopes with dried-out elastic bands around them. Swastikas, eagles and censorship stamps. Hundreds of them probably. Several had large, red postage stamps depicting a soldier wearing a German helmet and the words DEN NORSKE LEGION. A denomination of twenty plus eighty øre. Twenty øre for postage and eighty øre for the good fight. I skimmed through a couple of the letters from his fellow soldiers. Feldwebel Haraldsen thanked Sverre for his contribution.

Downstairs I heard Grubbe meowing. He padded around the house and came into the living room. Hopped up on the couch.

"He's dead, you know," I said, lifting him onto my lap and patting his belly. Grubbe was the only animal here now, the big forest cat with such long fur I worried the game committee would mistake him for a lynx. Before, we had hens and pigs and rabbits, but as I began to make more decisions about the running of the farm, I gradually cut down on the soft toys, as I called them.

I went back upstairs and went on searching. Came across a will from 1951. He had barely mentioned that year to me. Some sort of operation must have concerned him so much that he wanted to make his intentions clear: *All my possessions to Alma, the farm to Walter when he is of age. Would prefer cremation, if possible.*

He had never mentioned the last bit. Nor was it something that had come up in our chats at the kitchen table. All in all, the thought of Bestefar dying had been so remote. But this request was something Rannveig Landstad needed to know about.

I stood up and looked outside. It was half past twelve. I needed food and I needed cigarettes. The Texaco in Otta was all that was open in the valley at this ungodly hour. A forty-five-minute drive for two packs of Pall Mall and a microwave hamburger?

No. I might fall asleep behind the wheel, and besides, I wanted to be at the funeral parlour first thing in the morning.

It was time to sort out the difficult matters.

On an empty stomach, I slit open Mamma's envelope. In it was a gossamer-thin sheet of paper so brittle that it almost tore when I unfolded it. A birth certificate from March 1945. Issued in Malmö for a girl named Thérèse Maurel. The woman Mamma had borrowed the book from. Her birth certificate, here? And who was she, to have shown her birth certificate so many times that it was as thin as a cigarette paper?

The date of birth was given as January 15, 1945. The same birthday as my mother's. My head was muddled. I wondered whether she could have been Mamma's travelling companion, but deep down I knew that was not the case.

The next line revealed that Thérèse's mother was named Francine Maurel. Father unknown. The place of birth was Ravensbrück, Germany.

A child born in a death camp.

I felt a trembling unlike anything I had ever experienced before. Something tugging and pulling at my point of anchor. I searched for something fixed, something unalterable, and picked up my mother's passport. It had been cancelled after she died. One of the holes went through her photograph, through her cheek, but I could see both eyes.

The passport was issued in Paris in 1965. There it was, in black and white, my mother's name – Nicole Daireaux – with an address in Reims.

Reims? I had always thought she was from Authuille.

Mamma was photographed looking straight ahead. She had short hair and a deadly serious expression. Twenty years old. Why such a severe look when she was going on holiday to Norway, where she would end up meeting my father?

I met her eyes again as I glanced down at another document.

My hands were shaking. An unfamiliar guest had entered the room: the truth, in the form of a sheet of paper with three stamps. A certificate from a French registry office detailing a change of name.

Thérèse Maurel was indeed my mother's travelling companion. A constant shadow from the past. Just before the passport was issued, she had changed her name to Nicole Daireaux.

My mother was born in Ravensbrück, the concentration camp for women north of Berlin. Images swirled inside my head. Grainy black-and-white photographs of emaciated, half-naked people. *Father unknown*.

Until now my image of Mamma had been fixed. She had been a figure in blue, a goodness and a warmth that just *was*. She was part of a time, a chapter that had ended early, but it had been good.

But now her past had opened up and advanced its claims. All that was left in the envelope was a faded and creased identity card. A prisoner's card from Ravensbrück of a woman by the name of Isabelle Daireaux, from Authuille. Again that place. A magnetic field I could never escape. A burning vanishing point.

What had become of Isabelle? A prisoner's card like that was not something you left lying around. Either you burned it or you locked it deep inside a vault.

Or deep inside Mamma's envelope.

Mamma must have grown up without her parents, I realised. Like another person I met in the mirror every morning. Regardless, I now had a clue. Francine Maurel in Reims.

But something Hanne had once said stirred inside me. *You'll find nothing but old clues to torment you.*

I sat down with Bestefar's pile of letters. Censorship stamps and swastikas. I felt the urge to burn everything, to go to the fields and tend the potatoes. When was I going to discover who *I* was, the *real me*? It was as though everything inside me flowed into an enormous pool of water, with a film of soldier's blood and used gun oil on the surface. A membrane so thick that if I did not manage to force myself up through it, I would drown.

I found an out-of-focus photograph, streaked with green. My mother and Bestefar on the stone steps of the cottage. They did not seem to know they were being photographed. Mamma was wearing a kerchief and she was terribly thin, almost emaciated.

I held it up to the lamp, noticing that a piece of paper was glued to the back. I slipped in the knife and Alma's handwriting appeared.

The French . . .

The paper tore and snagged in the glue. I tried from the other side. The fragments of paper resembled stubborn heaps of snow in the spring. Carefully I scraped them off with my nail.

The French drifter. April 1966.

What had she meant by it? That Mamma was some kind of fortune hunter?

Someone had glued paper over the comment. Bestefar? Or had Alma had second thoughts?

I looked again in the envelope but found nothing more about Mamma's past. Just copies of Bestefar's letter to the police chief in Saksum, in which he referred to "The Act concerning the entry of foreign nationals into the Kingdom."

"Nicole Daireaux continues to live and work here at Hirifjell and is in no way a strain on the public purse, thus the Immigration Act gives her the right to reside in Norway, also in future years."

I took the magnifying glass and studied the photograph again. Her clothes were grey and shabby. Her hair was straggly beneath her kerchief and she was clinging to a bulging plastic bag.

She was much skinnier than in the passport photograph. Who was she, arriving with her clothes in a plastic bag from a French grocery store? As far as I was aware, Alma barely knew which way to hold up a camera. Yet had she been the one who had photographed Mamma, in secret? Or had Pappa taken the picture?

No, he had been working in Oslo at the time. Alma would not

have called Mamma a drifter if she had met Pappa in Oslo first and come with him. She would not have stood like that either, as if she had been walking along a railway line for days. The only explanation could be that she had come to the farm *before* she met Pappa.

That raised a bigger question.

Why would a French girl, an impoverished adoptee, seek out a remote farm in the mountains of Norway?

4

I WASHED THE STAR WITH THE HIGH-PRESSURE HOSE and drove to Saksum. It was half past eight and probably too early, but all my life it had been impossible to tell whether H. Landstad and Family was open or closed. No movements had ever been discernible behind the curtains, not that I had paid much attention. I had shunned the silent unpleasantness of the funeral parlour like the plague.

The door was locked. I got back into the car and studied the documents in the glove compartment while I waited. Punctual service stamps from Lillehammer Motorcentral. How much could the car be worth? A black S-Class driven less than four thousand klicks a year. With one exception. In 1971 it had travelled nine thousand kilometres.

I know that Sverre held Nicole dear, the old priest had said.

In fact so dear that it had made its mark in the service logbook. Bestefar had known their travel plans, lent them the new car.

I turned and looked at the back seat. There I had sat. The night before, I had found Bestefar's aeroplane ticket to France, one way. The Star's registration number was listed on a ferry ticket dated four days later. On the trip home, only the two of us.

I shut my eyes and tried to conjure up memories of those four days, but nothing came. Sometimes I had a feeling that something terrifying had happened in a car, a hysterical voice with the smell of exhaust and old leather, but it could not have been this car. The Mercedes, with its synthetic leather seats and the sombre humming of the motor, I associated only with something safe. If I remembered correctly, that is.

Inside the funeral parlour, the lights came on.

No doorbell. My footsteps were deadened by a dark carpet. The light was monotone and subdued, perhaps a quarter of a second on a 2.8 stop. Four chairs around a black table. Apparently there was little need for furniture when you had a monopoly on death in the village.

She emerged from the back room wearing dark-grey office clothes. Walked around the counter, shook my hand and did not let go. She said nothing, but gave me to understand that I had been expected. At first I thought that this was a silence meant for all types of bereaved: parents who were broken for life, having had to order a small coffin; the spouses of tyrants who were happy that the bastard was finally gone. But Rannveig Landstad's silence flowed inside me like a sedative, and all at once – for the first time in ages – it felt as if I had something in common with others in the village. Other people had stood here and felt the same way, stood here shaken and destroyed in the antechamber of the churchyard, and I was not ashamed that I was red-eyed and out of sorts having wandered around the entire night burrowing through papers before lying sleepless, staring at the clock.

She released my hand before it got clammy and asked me to sit. She lifted a leather portfolio, slipped a ruled sheet of paper under the clip and pressed the end of a gold-plated pen.

"The coffin," I said.

She looked unsure. Clicked the pen again.

"The priest told me," I said, "that someone sent my grand-father a coffin."

"Yes. It . . . there is a coffin here. Or rather. Of course there are. Coffins here, I mean. What I mean is that I cannot remember us ever having this kind of, well, arrangement. But I suggest that we take care of practical matters first."

Then she was back on track, practised. She began with the

simple things, so that those in mourning would not keel over and find the task insurmountable. She nodded as she jotted down the request for cremation. The gravestone was also straightforward; we were a frugal and prudent family, we had left space on Alma's stone. It was the same as Mamma's and Pappa's, a greyish-blue Saksum granite found only on a mountain knoll down by Laugen, beneath the railway bridge.

"Flowers," I said. "There should be flowers around the coffin, right?"

"Absolutely. We will also place wreaths around it, from friends and family."

"There is hardly any family to speak of," I said. Someone from Alma's family in Ringebu would presumably send a wreath, and that was it. It was unlikely that the Sheep and Goat Breeders would send anything to an inactive member in Saksum.

Rannveig waited a few seconds, twiddled the pen in her hand. "We can organise some lovely flowers. Jarle's do good work. When there's a well-chosen arrangement, with nice colours, it doesn't matter if it looks a little sparse."

"It's not going to look sparse," I said. "What do you think of having potato flowers around the coffin?"

"*Potato* flowers?"

"There are loads of them right now, I could fill the boot. Red and violet flowers from the pimpernel, and white from the almond potatoes."

Rannveig Landstad shifted her grip on the pen. "I see nothing wrong with that. In fact, that could be quite good." She paused. "Do you have anyone staying with you? Anyone . . . close?"

Had the nosiness of the village even worked its way in here? Was she curious about me and Hanne?

"I have friends over," I said.

"Keep them close. It will be difficult for you to face this alone. The next few days will probably be the worst."

Suddenly I longed to return to the practical matters, to talk about Bestefar, and not about Hanne. Because I remembered that she *charged* for this gentle voice, it was all included in the price, and when she had done her job and Bestefar was lying in his grave, she would no longer be paid to comfort me.

Rannveig Landstad twirled the pen in her hand. Small ripples ran down her freshly ironed blouse.

"In the spring of 1979," she said, "a vehicle arrived from a freight company, loaded with a large wooden crate. Inside, wrapped in sailcloth, was a coffin. Tied to one of the handles was an envelope. We found a letter and a sum of money to pay for storage. It was all rather ... unusual."

"But why didn't you tell my grandfather about it?"

"The letter specified that he wasn't to know about the coffin. That *you* should decide if it was to be used."

"Me? Was someone trying to taunt him?"

"No, no. Heavens, no. Then we would never have accepted the request. Dear God, no. There is nothing tactless about this. On the contrary, it's rather an extraordinary coffin. Without disparaging the others we've supplied over the years, this is the finest coffin anyone from Saksum has ever been buried in. It would befit a state funeral."

"It was his brother," I said. "Einar. I thought he was dead."

She looked up at me. "My apologies if this has made the occasion even more difficult," she said.

I released the air from my lungs. "Where's the coffin now?"

"In the storeroom. I removed the sailcloth earlier this morning. To tell you the truth, it will be nice to settle this matter once and for all."

"Do you have the letter?" I said.

She must have placed it inside the leather portfolio that morning. It was typewritten, single-spaced, and she handed it to me with the same look Bestefar got when we played cards and I was about to draw the Old Maid.

Coffin for Sverre Hirifjell. He has no knowledge of the gift and should not be made aware of it. Upon Sverre's death,

Edvard is to decide whether or not the coffin should be used. In the tragic event that Edvard dies before Sverre, I request that Edvard be given his final rest in it. Should this be the case, this letter can be shown to Sverre. If the coffin is not put to use, it is to be burned. With no-one present but the undertakers. The coffin is not to be painted or varnished. Fire or earth, nothing else.

"Did you show the letter to the old priest?" I said.

"No, there are limits to what we would share even with him. But of course we have maintained a certain level of . . . cooperation over the years. He used to invite himself round for coffee every other day. When the coffin arrived, he studied it carefully."

"And?"

"And what?"

"How did he react?"

"Like you, he said, 'It must be Einar. Only he could make something like that.'"

I followed Rannveig Landstad to the end of a corridor. The storeroom was chilly and smelled of concrete and stone. Old ring binders and a crate of tarnished silver-plated candlesticks. On deep shelves, two rows along each wall, were the coffins. Most with a white gloss finish, some made of pine, a couple of black ones. Leaning against the wall were the headstone samples. Like left luggage on the platform of the dead.

"The old priest is going to officiate," I said as we walked through the storeroom. "The new priest is apparently on holiday."

"Holiday?" Rannveig Landstad said, opening a door.

"That's what he said. That the new priest was entitled to a holiday."

"That may be. But he's just got back from Rhodes."

"Really?"

"Yes. I believe Thallaug sees it as his sacred duty to take care of your family." She switched on a ceiling lamp, grabbed my arm and turned me in the right direction.

The coffin was resting on a large table covered with white fabric that reached the floor. I stood gaping.

The style was one thing; it was polygonal and had infinite facets that flickered in the light. But it was the woodwork that really bowled me over. The amber flame-birch scintillated. In the dimly lit room it was practically luminous. Over the intense base colour, yellowy-orange tongues twisted in long and unpredictable patterns. Dense structures changed shape, leaped out like claws, and were different from whichever angle you looked. Carved into the lid was a barely visible check pattern, embellishments that made the light and shadows continually give off new shades of colours and radiance.

I approached it. The angles of the wood were sharp enough to cut yourself on. The lid was so precise a fit that it was impossible to find the seam. At first glance I thought it was varnished. But the wood was waxed and *polished*.

The spring of 1979. The year after Bestefar refused Einar permission to see me on my tenth birthday. In response he felled four trees in the flame-birch woods. Enough for a coffin.

But it was no offer of reconciliation, I told myself. The coffin was a message. A precise, time-activated message. One that would reach me as soon as Bestefar was dead.

"An *art-deco* coffin," Rannveig Landstad said. "Imagine that."

I stared at her for a long time.

"Is there such a thing as an *art-deco* coffin?" I said.

"I suppose this is proof of that."

"Have you ever opened it?"

"Even we are human," Rannveig Landstad said, placing her fingers against one of the grooves. A horizontal opening grew to a black chasm. Effortlessly, she lifted the lid all the way up. Two chrome-plated balance springs held it in equilibrium. A shiny brass piano hinge was sunk into the entire length of the coffin, and I noticed that the hollow of every wood screw lined up, a level of perfection our woodwork teacher had exhorted us to aspire to.

The coffin was not lined with velvet, as I had at first thought. It was veneered with the same kind of wood as my shotgun. Similar to flame-birch, but even wilder, more unruly.

Like the fiery glow of hell. Or flowers bent over during a storm.

A mild morning. I woke up on the couch, fully clothed and sweating. Went into the kitchen and studied the papers I had gathered from Bestefar's desk. Before collapsing the night before, I had sorted everything, arranged page after page across the floor until they covered the entire drawing room, but all I found were old letters from the agricultural control office and insurance contracts from Gjensidige.

He had been honest, my grandfather. Everything pertaining to the family was gathered in those envelopes. With one possible exception. In the corner of a drawer I had found a small bunch of keys. Three shiny padlock keys from O. MUSTAD & SON along with a weathered, wrought-iron key. The key ring was attached to a rectangular piece of wood. The keys might have been for a lock up at the mountain pasture, or for a boat, but something made me inspect the wood more closely. It was scored and reddish-brown, and when I held it under the light I recognised the pattern. It was walnut.

In Pappa's envelope I had found only school reports and certificates. Pappa in numbers and letters, as he would have been seen by a tax official. As he was still seen by me.

Apart from Mamma, the papers breathed life only into Einar, despite there being only three items in his envelope.

A telegram from Paris dated July 12, 1938. MY BROTHER. RECEIVED NEWS AFT 1 MONTH AWAY FROM APTMENT. GRIEVING FOR FATHER. PLEASE LEAVE FLOWERS ON GRAVE. EINAR.

A tawny photograph of what must be him, standing next to an enormous country dresser with intricate carvings. He looked like the photograph of Bestefar on his N.S. membership card, but was slighter and had a peculiar half smile, as though he had been asked an unexpected question.

A form. *Notice of death to the parish priest from the Probate Court. Full name of the deceased: Einar Hirifjell. Died: The night of 2 to 3 February, 1944. Place: Authuille, France.* The field "burial or cremation" had been crossed out.

Attached to the form with a rusty paper clip was the German death certificate, no bigger than a fishing licence. Point 5 gave the cause of death: *Hingerichtet*. Executed. Stamped with a German eagle and a swastika.

I rounded the storehouse and walked over to the carpentry workshop. It had always been like this; completely isolated, with faded red paint, filthy windows and roof tiles covered with moss, as though it stood alone, with its own thoughts. The key had hung in a cupboard in the log house for all these years, accessible, but nonetheless forbidden. Once as a small child I had let myself in, but did not care for the darkness within, the hazy outlines of tools and materials. The layer of dust was so thick I thought the floor was carpeted.

Now the door was swollen, but yielded to a firm kick. I stood in the doorway. The workshop smelled withered. The windows were a greasy, yellowish-brown.

On the floor I could make out my footprints from when I was small. Much smaller shoes. There was a furrow in the dust on the lathe. I must have run my index finger along it and drawn out the underlying colour. There were no other signs of activity. If Einar had returned to the farm in 1978, he had not been in here.

I fetched a work light from the tool shed and unwound an extension cord in the grass as I walked back. The light gave the room purpose. The carpenter's bench, hand tools on the wall, materials in the rafters, a half-finished chair in one corner. The layer of dust made everything resemble a sepia photograph.

I took two bottles from the windowsill and brushed off the labels. Linseed oil no. 8. Shellac no. 2. The solvent had evaporated and their contents had long since dried into bone-coloured layers, paint remnants in lines down the glass. The dust made me sneeze, my movements swirled up more dust and I sneezed again and had to force myself to stand still. Colours appeared everywhere I looked.

The departure for the Shetland Islands had not been hurried. The workshop was tidy, there were no wood shavings in the nooks of the workbench, the tools all hung in their place.

Then I looked outside at what had been Einar's view.

At first I had thought that the layer of dust on the windows made it impossible to see *in*, but now I realised that it was *the farm* which had become blurred behind the caked windowpanes.

I fetched a dust mask and broom and swept out the worst of the mess. Washed the windowpanes and positioned a ladder next to the electricity intake below the eaves. The cables had been severed and pointed skywards like crooked fingers. I stripped the extension wire and hooked up the power. Through the window I saw the lights come on.

There were tools inside that we could have used on the farm. A bandsaw with a dark-green hammer finish. An electric planer. A complete set of hand tools. Wood chisels, screwdrivers, saws. The lathe rumbled to life when I switched it on, an enormous mechanism with a heavy fabric strap and a flywheel worn smooth, powdery old wood chips caked in dried grease. A gentle nudge to the spindle, a singed smell for a few seconds, then it spun into action and whistled through the air.

Once, when I was playing, I had broken the leg of one of the nice chairs. I said that I could fix it if only I had a lathe. Bestefar said there was no need for a lathe, and an hour later he had swapped the round, straight-lathed chair legs with two rough, square ones.

That was Bestefar's idea of carpentry. Nail heads poking up. A little *too* sturdy. As though he was trying to escape from somebody's shadow.

*

In a cupboard there was a row of old books. *L'Art du menuisier ébéniste. Anatomie du meuble*. Working drawings of furniture, elaborate constructions with a myriad of details. Chests with forty small drawers. A round cupboard with sliding doors.

The most worn publication was a catalogue from a furniture exhibition in Paris in 1925. It was in French, and my stomach fluttered when I realised I understood most of it. On the front cover was a drawing of a girl – or was she a faun? – wearing a loose-fitting dress and carrying a basket of flowers, running with an antelope across a meadow.

I sat down in the heat of the rising sun and read. A crow cawed and flapped off towards the woods.

Here he had sat. Right *here*. Perhaps had heard the crows taking off from the pine wood he was condemned to build plain wooden furniture with. Birds from the same family that were here now. He had dreamed of this, magnificent furniture of a style I had never seen before, and over the course of my life I might never see, let alone meet someone who owned anything like it. The designs, the decoration and the patterns surpassed one another with each page. And in addition there was someone attempting to outdo them; on every centimetre of free space, Einar had added his own ideas. He had set aside a style that was already audacious, shaded in another type of wood, changed the pattern on the frosted-glass doors, swapped in some etched tulips with an intricate check pattern.

On a loose sheet of paper were his plans for the flame-birch woods. The distance between the trunks. How the withies were to be moved. *Loosen A, D and E every other year. B and C every fifth year.*

I opened *Anatomie du meuble*. On the first blank page was written "Einar Hirifjell, Paris 1933". Such handwriting he had, upright and straight, and with a cross stroke on the "H" that ran through the length of his surname. An "H" which was also mine. My surname, which previously I had heard tainted in war stories, stood here proudly, and painstakingly made.

We could have been a family. We could have had Christmas parties enveloped in cigar smoke, with tales of great voyages. We could have played under the table, tugged at the shirts of the grown-ups who stood at the window waiting for out-of-county cars to arrive in the farmyard.

What would it take for me to be able to write "Hirifjell" with such pride?

I put the book down. Bestefar's reality was to wake up each day and till the soil. Why would I not be like him?

I savoured the possibility. That the envelopes in the desk did not exist. That no coffin had arrived from Shetland. That I could continue to see the meadow spring up, drive out every morning in the Deutz which I had hosed down the previous evening.

Perhaps lies are like alcohol, I thought. You have to drink regularly to hide from yourself the fact that you drink. But perhaps there is a similarity with the truth. You have to drink until the bottle is empty.

"I'm sorry, but that name is not in the directory. There is no Einar Hirifjell in the Shetland Islands."

The answers from Televerket's international enquiry service came calm and prepared.

"Of course," said the lady at the other end, "that does not mean the person does not exist. Or didn't."

"He may have had a post office box in Lerwick," I said. "Number 118. Does that help?"

"When would that have been?" the lady said. She had an Oslo accent, with a hint of Trøndelag dialect.

"1967."

"That's more than twenty years ago," she said without sarcasm. I imagined there was a hierarchy where she worked, something like the distinction between the amateur and professional departments at the camera shop in Oslo.

"Let me investigate," she said. "I'll call you back. It may take some time."

I replaced the receiver and picked up the photograph of Mamma and Pappa; I should move either the picture or the telephone to the cottage. I sat on the stairs and opened a small photograph album I had found in the carpentry workshop. Street scenes from Paris.

A ticket to "Nosferatu" at Le Grand Rex. A vast cinema, apparently; he had sat in seat 48, row 60.

A photograph of an immense workshop, four men in work clothes standing around a colossal cabinet. A close-up of a young man wearing an overall coat, putting on airs and holding up a wood chisel. *Charles B.* The next image was of a man wearing round spectacles, his hair parted down the middle. *Ruhlmann.* On his desk was a set square and sketches of a divan.

The photographs on the next page showed two men in sweat-stained shirts sitting on a raft. *Bonsergent and E. Hirifjell in Gabon, 1938. Concluded agreement with Lacroix for annual delivery of 30 m³ of bubinga wood.*

A small sketchbook dated 1926. How old would he have been? Twelve? Even then he had been drawing street scenes and splendid houses, as well as living spaces with magnificent furniture. On the back cover he had written a checklist:

> *– Clean the barn before they have to ask.*
> *– Ignore Sverre when he goads you.*
> *– Spend 30 min. practising calligraphy and technical writing.*
> *– Help out if Mamma is left with the heavy work.*
> *– Carve something elaborate freehand.*
> *– Spend at least an hour practising dovetailing or bridle joints.*
> *– Be more polite at the table. Clear up.*
> *– Suggest repairs at the farm before I make something new.*

I pictured the two of us at the farm. What it would have been like if *we* had been brothers. An hour later the telephone rang, harsh and metallic.

"This is Regine Anderson from Televerket's international

enquiry service returning your call. My apologies that it took so long. Now, are you *certain* that Lerwick 118 has a connection to the man you're looking for?"

"Quite certain."

"The problem is, this cannot possibly have been his address. In 1967 there were only eighty P.O. boxes in Lerwick. Most letters were delivered according to name and street number. But I did discover something else. Lerwick 118 was the telephone number of a hairdresser's on St Sunniva Street."

"A hairdresser's?"

"Yes. St Sunniva Hairdresser's. Near the crossing with King Haakon Street. I've studied the map."

"Hm," I said faintly. "Well, thank you for checking."

At that moment Regine Anderson paused dramatically, a prelude to a performance that must have compensated for days filled with sullen, ungrateful enquiries.

"I sensed that this was important to you," she said. "So I had a colleague at British Telecom in Aberdeen search the archives. He discovered something interesting."

"I see."

"The hairdresser's had this number since 1937. But for twenty-one years, from 1946 up to and including 1967, the directory for Shetland had *two* entries for Lerwick 118. One was St Sunniva Hairdresser's. The other was a certain E. Hirifjell."

I felt a tingling in my stomach. "So if you looked up under H in the Shetland directory, he was listed with that number?"

I heard paper shuffling at the other end, notes from routine fourteen-digit enquiries for international numbers. "Yes," she said, "but not after 1968."

Alma must have tracked him down in 1967, I thought. By the following year, when I was born, he had removed his entry from the directory.

"Are you still there?" Regine Anderson said.

"Yes. Yes of course. Was the hairdresser's number removed at the same time?"

"No, it was in use until 1975. Then something interesting happened. Old numbers are normally placed in a three-year quarantine before they are reused, because many people memorise them and call, thinking the company is still operating. Unusually, this number never became inactive – it was transferred to a lady by the name of Agnes Brown on St Sunniva Street. She probably lives above the old hairdresser's."

"She must have been the owner," I said, imagining that she might have been married to Einar, and that he had later moved out.

"Most likely. Because this Agnes Brown is still in this year's directory. Same address, same number, just with another dialling code after the switchboard was automated. It's odd that she hasn't changed her number. As a pensioner, you would want to avoid being woken up morning, noon and night, wouldn't you?"

"Maybe she's expecting someone to call," I said.

"That's generally why people have telephones," Regine Anderson said. I could hear that she found that amusing. "Would you like the number?"

"Absolutely," I said. "You would have made a great detective."

"That I am, young man. I've been working for Televerket's international enquiry service for thirty-nine years."

It was the first time I had dialled an international number. There was a grainy whirring on the line, as though the signals were fighting their way under the North Sea. It rang at the other end, a loud jangling that sounded different from Norwegian telephones.

I held the receiver, waited.

No answer.

I hung up, paced the room and thought about the telephone number. I went downstairs to Bestefar's atlas and studied the distances. I knew that the Shetland Islands had once been Norwegian, and now I discovered why; the distance to Lerwick from Bergen was shorter than from Aberdeen.

Then the telephone jingled again. I leaped up the stairs and snatched the receiver.

"Yes," I said in English. "Hello?"

"What?" the voice at the other end said.

"Hm?"

"This is Rannveig Landstad. Is that you, Edvard? Listen, I'm sorry to have to tell you, but some complications have arisen."

She told me that the coffin could not be used for cremation. It was too wide for the furnace. Presumably it had been made according to English measurements, because it was exactly four feet wide.

No enlightened nation measures in feet, Bestefar had said.

"By the way, did you intend for his knife to be buried with him?" she said.

"He would have appreciated that."

"We probably could have left it in the coffin, even though it's not entirely by the book. But with cremation – well, you understand."

I walked around the farm for an hour, deep in thought. Wandered among the berry bushes, filled up on them and looked back across the fields at the carpentry workshop.

Fire or earth. Nothing else.

I called the funeral parlour.

"Can we use the coffin for the church service?" I said. "And take him out before the cremation? Would that work?"

"But then," Rannveig Landstad said, "then we are left with a used coffin."

"You could drive it up here to Hirifjell. Leave the knife inside."

"Edvard, it is unusual in my trade to ask *why*, but today I believe it is necessary."

"There'll be a need for the coffin later," I said.

I knew that he lived in the blind alley near the agricultural co-operative, in a single-storey black house overshadowed with bushy spruce trees. They were probably just Christmas trees when he moved in; now they had taken over the entire lot. The felt roofing was overgrown with moss, the gutters filled with spruce needles. The Rover was in the carport, but there was no answer when I rang the bell. I went round the side of the house and found him in the back garden.

"Who was Thérèse Maurel?" I said.

He shifted in the sunken folding chair. "Your mother was your mother," he said, and pointed to another chair leaning against the drainpipe. I unfolded it and sat down in front of him. He was drinking from a one-litre bottle and offered it to me. Out of politeness, I did not wipe the spout before drinking. Apple juice, thick and sweet.

"Why didn't you tell me everything at once?" I said.

"Sorrow is most pure when it has a fixed point. I thought it might come out later. Funerals for the Hirifjell family have never been simple. The coffin you would have found out about regardless, but I wondered what you knew and did not know about your past. And how much you *wanted* to know. Sometimes the truth needs to bide its time."

The grass brushed against the flowered seat covers we were sitting on.

"I think the time has come," I said.

"Your grandmother was named Isabelle Daireaux. She gave birth to your mother in Ravensbrück concentration camp, just before the capitulation. They must have been separated from each other, because your mother was adopted by a French woman, Francine Maurel, and grew up believing she was her real mother. When she was seventeen, she uncovered documents that revealed this was not the case. At the age of twenty she changed her name back to Daireaux, to Nicole Daireaux."

"Was it Sverre who told you this?"

"No, it was your father. I had to write out your Norwegian birth certificate, and needed identification papers and personal identity numbers. Then the family history came to light. I saw her documents."

"What happened to her? My real grandmother?"

"I was never told. But if she survived a camp like that, and lost a child, she could not possibly have escaped from there without great suffering."

I told the priest about the papers in the secretary desk and the photograph of Mamma. "What I don't understand," I said, "is why my mother came *here*, of all places?"

"That is a secret they kept to themselves. Walter insisted that she was a tourist."

"In that case, she was the first person ever to choose the far side of Saksum as a destination."

"Don't underestimate the rural community. Bethlehem was no metropolis either."

I looked for a stone or something else to fiddle with, but there was nothing like that around. I realised I was sitting with my hands tightly folded, and relaxed them.

"I was wondering why she wanted to be called Nicole," I said. "Go on."

"Mamma's birth certificate indicated that the father's name was unknown. Did she know anything more by the time you met her?"

"Not that I was told, unfortunately."

"You *are* telling me the truth now, yes?"

"I always speak the truth. It is just that sometimes I choose not to speak."

"I went to see the coffin Einar sent," I said. "I think he came here on my tenth birthday and chopped down the trees it was made from."

The priest looked across the unmown lawn. "Did he so?" he said in surprise. "All I know is that Einar was back in Saksum the year before you were born. Here, have a little more apple juice. Yes, I certainly meant to tell you when the time came."

Sometime during the summer of 1967, Einar had appeared in the doorway of the parsonage. Magnus Thallaug was sitting there with a newspaper and a coffee, believing that Einar had died in 1944. Had he not spoken with a Gudbrandsdalen dialect, Thallaug would never have recognised him, because Einar's face was as furrowed as the bark of a holly tree. The lively and healthy young man who had repaired the crucifix and the altarpiece of Saksum church now stood ragged and trembling. Far too old for his age, his body looked like a cowhide hung over a post. The priest had seen suffering in him. Years without a good night's sleep, unvaried food and too little soap. The only thing distinguishing Einar from a vagrant was his well-groomed hair.

In the end he disclosed to the priest that he had been living alone in Shetland since the war. "I received a report that you had been killed," the priest said.

"I wish I had been," Einar said. "But I had to meet the girl who has come to Hirifjell."

There was a grey car with English number plates in the courtyard, as dented and disfigured as Einar himself. The priest had thought Einar was in Saksum to visit his home town one last time, and asked him to repeat his request. Yes, it *was* Nicole he was looking for. He had been out to Hirifjell, but left because Nicole – my mother – had apparently been furious with him. The priest had asked why, and Einar replied that it was a matter between himself and God.

"In which case I am the very man to help," the priest said. "With me you're as close to Him as you will get without being dead. Now, tell me how I can help."

Einar stood there irresolute. He had evidently not devised a plan, had just stopped by the priest's office as if it were a lighthouse, where clever ideas came spontaneously.

"Give me a pen and a piece of paper," Einar said after a while. "I'll write to her."

In Einar, the priest recognised something that was seldom seen in the village of Saksum. He had become religious. But the faith that filled him was not the sing-songy kind, with baskets of flowers by the door. His doctrine was rock-hard and filled with remorse and anguish. Yet Einar refused to speak of the desperation that plagued him.

The parish council was meeting that day, and the priest offered him something from the tray of sandwiches before he went to put on more coffee. When he returned with the fresh pot, Einar had devoured everything, as though he had seen no food for a week. The priest considered this entirely possible. Einar settled into the priest's office, wrote the letter and was gone.

As a priest in Gudbrandsdalen, you see a bit of everything, thought Thallaug. But this incident had given him, in his own words, "a need to see if the Church ought not contribute a little extra to ease the poverty of its parishioners". So the following day he had taken the Rover and driven out to Hirifjell. The farm was apparently deserted. Not a soul in sight. But the priest heard voices coming from the kitchen garden. Mamma and Einar were standing beneath a plum tree. They were speaking in French, and they were speaking loudly, in agitated voices, but it was no argument. When the priest arrived, they went silent. Mamma curtsied to him and exchanged a few polite words in Norwegian, and then she went inside the cottage. The priest strolled around the farm with Einar. He said that in the end Mamma had "realised what was in her best interests", but the priest was none the wiser. The discord between the brothers seemed to persist, because Sverre and Alma had travelled into town with Walter.

The priest did not learn much about Einar's life, other than that he had settled on the Shetland Islands. It was unclear where, exactly, as the only two place names Einar muttered were Scalloway and Unst. Scalloway was familiar to the priest, its harbour had been the wartime base for the clandestine Shetland Bus, but later the priest had to check the atlas to discover that Unst was the northernmost island, a desolate and almost deserted place. Thallaug had hinted that Shetland was a strange place for a man who had been Ruhlmann's master cabinetmaker, but he did not get much out of Einar about his life after 1942, nor about his connection to Mamma or why he had to write to her before she would speak to him.

Einar was remote and somehow strange, and the conversation soon grew as stilted and reluctant as it had at the priest's office. But the priest believed he had seen something between Einar and Nicole, something unresolved. The priest left Einar standing on his own in the farmyard, looking up into the woods. Then Mamma emerged again from the cottage.

"And you were going to keep all this to yourself?"

The priest took off his glasses and rubbed his eyes.

"I thought you needed the truth in small portions," he said, putting his glasses back on. "No-one jumps onto a wobbly ladder. But now I'll tell you absolutely everything I know."

I plucked a blade of grass. "You said something about his hair. That he was well groomed."

"Yes, it was something I noticed. Apart from that he was rather worn and frail. Wore ugly yellow wellies. The last time I had seen him, when he came back from Paris in the thirties, he was pompous and gallant."

I told the priest about Agnes Brown and St Sunniva Hairdresser's.

"What do you think his conflict with Bestefar was about?" I said. "It was more than the war, wasn't it?"

The priest emptied the bottle of juice.

"First let me ask you about the four days in France," the priest said. "Do they torment you because you are *afraid* of what happened, or because you don't *know* what happened?"

"Is there a difference?"

"Oh, yes. Some people survive best by nailing together some kind of truth. Even if it's crooked or full of cracks, it might still hold. For many people it blocks out life."

"I want you to tell me everything you know."

"To begin with, the feud between the Hirifjell brothers was most probably about allodial rights and politics. I believe the dispute was renewed after the accident in 1971, and the tone must have changed. Because blood is thicker than water, and that is at the heart of the matter. Forgive me for speculating, but I think Einar knew what happened to your parents, he just wouldn't talk about it."

A twitch travelled across my forehead and lodged itself in my eyelids, then spread downwards until it had taken hold of my entire body. My stomach knotted up until something seemed to snap and sorrow coloured every new thought I had. What a foolish curiosity I had succumbed to. It was as if I had taken apart something expensive just for the hell of it, only to realise I couldn't put it back together again.

The priest straightened up.

"You asked me to tell you everything, Edvard. That is how it feels. And I am not finished yet. You have to accept it: you have to transfer all the stones into your rucksack."

"But what," I muttered, "what makes you think Einar knew?"

"He didn't come to the funeral. Either he was not wanted there or he could not bring himself to come. Both would suggest that he was involved, because previously he had been obsessed with meeting your mother."

"Maybe nobody told him," I said.

"I don't think he *needed* to be told," the priest said, and then described how Bestefar, normally as quiet as salt lick, had requested that Mamma and Pappa share a coffin and a head-stone, even though they were not married. When finally they were lowered into the ground, both he and Alma fell to their knees bawling.

"It was a good, healthy reaction, in fact. But I heard Sverre repeating Einar's name as he sobbed into the ground. He said something about 'those damned woods', mumbled it over and over. Anger and compassion mixed together. As though he wanted to both punish his brother and accept him."

"Did he really say 'those damned woods'?"

"Several times."

"Did he mean the flame-birch woods?"

"No. It sounded as though he was referring to the place where they died, where you disappeared."

I got up and walked towards the fence. The twitching in my eyelids had stopped, but I knew that from now on, nothing would be the same again.

"Did he not come to the churchyard later?" I said. "Einar, I mean."

"No. I kept an eye on the grave, but only one person ever visited it."

"Bestefar?"

The priest shook his head. "He wasn't made like that, you know. The snow arrived early in 1971, and throughout the winter only a narrow set of footprints led to their grave – Alma's."

"That was how Sverre dealt with things," I said. "Took it out on his work."

"And you?" the priest said. "How are you dealing with it?"

I swallowed. Realised there are very few turning points like this, when you stare up at the clouds and promise yourself that everything will be different from now on. But even the most stringent resolutions loosen their grip over time, so the oath must be sworn while it still hurts. In my head and out of habit, I steered towards Bestefar, towards me becoming the salt lick. But my body wanted something else. It wanted breakdowns and tears, lightning bolts and reckless acts, if only to prove that I was not numbed and hardened. Because I realised that what I longed for most of all was to feel real loss.

A minute went by, maybe ten. I stood by the fence. The old priest still sat in the folding chair, staring at me as if I were a beloved farm animal ripe for the slaughter. As if reluctantly assessing how much I could take before I toppled.

"There is only one stone left now," he said.

"Out with it," I said, and ambled towards him.

"The problem is that this is the weightiest and roughest of

all. It concerns an unresolved matter relating to your mother."

"Unresolved?"

"I mentioned that Einar wrote a letter to Nicole in my office. He used the parish newsletter as a blotter. At the time, we were splashing out on thick, glossy paper. When I tidied up after him, I saw that he had pressed so hard that some words were visible on the paper beneath."

The priest heaved himself up out of his chair and I followed him into the house, through a musty kitchen and into a narrow office. Its four walls were filled with bookshelves, and typewritten pages with a multitude of corrections and notes poked out from between book spines and archive folders. He got down stiffly on one knee and pulled out a brown file. Inside it was an old parish newsletter.

"I have kept it here," he said. "In the unlikely event that some- one from the Hirifjell family might want to plough deep into the past."

The sunlight entered the room at a slant, casting tiny shadows on the letters. The paper fibres had straightened over the years, but when I held the sheet flat up to the window, I recognised faint traces of Einar's meticulous handwriting. The lines crossed one another and words blotted out others, but some were still legible. On a blank area beneath the drawing of Saksum church I saw two names. *Oscar Ribaut*, with the year 1944 written next to it, and *Isabelle Daireaux*.

"Who is Ribaut?" I asked.

"I don't know," the priest said, pulling a hair from his nostril. "I may be punished severely for telling you this – because now I'm opening a gate your grandfather kept locked for twenty years, presumably with good reason – but if you look closely you'll see one word repeated three times, there, there and . . . there."

I followed the yellowed nail of his index finger across the faint impressions from Einar's pencil.

The repeated word was "*l'héritage*". Something to do with an inheritance.

"Whose?" I said.

The old priest cleared his throat to make me understand that, in this matter, he may well have been a little too tenacious in his role as spiritual advisor. On that day in 1967 he had sprinkled ashes over the parish newsletter and tried to decipher Einar's message. Much to his surprise he had discovered that it was about an inheritance belonging to the Daireaux family. It seemed to be extremely valuable, either in money terms or because of its sentimental value, and the priest gathered that it was old, several hundred years old. But the words were laid over each other and it was not clear whether Mamma knew where the inheritance was, or whether Einar believed that she was its rightful heir. In any case, the priest realised that Mamma had intended to settle down on the farm, but Einar had awoken something inside her.

"I think Einar persuaded them to take the ill-fated journey to France in 1971," the priest said, placing the parish newsletter back in the folder. "What this inheritance might have been, I do not know. But I overheard something Einar said to your mother under the plum tree. My comprehension of French is not exceptional, but it agreed with the sentences I had read on the sheet of paper. Einar said that the entire inheritance still existed, and that there was enough to fill a lorry."

"So we were going back to the farm in Authuille? Where the Daireaux name originates?"

"It would seem so."

"Do you think Einar is still alive?"

"Quite possibly. I'm still alive, after all. His body was wasted, but I saw an obsession so ardent that it could keep a man alive for a hundred years. He would not yet be eighty."

"I still can't understand why my mother came to Hirifjell, of all places."

"Nor I. But I must send you away with one caution. Your mother's surname was originally Maurel. Whether she changed her name in order to be taken for the true heir, or whether she became aware of the inheritance only through Einar, that I don't know."

Back home, a white Manta was parked on the unmowed grass. The bonnet was warm. She sat in the conservatory reading a text-book on feed concentrate for piglets. She wore ripped jeans and was scratching a mosquito bite on her tanned thigh. Grubbe was lying on a pillow with one paw over his nose.

"Would you like some blackcurrants," she said, and nodded at the extra bowl and a one-litre jug filled with berries.

"Sounds good," I said.

"The key was in the usual place," she said.

"And now you're the only other person who knows where it is," I said, hanging up my denim jacket. I took in the sight of her, the possibility of Hanne Solvoll. For this summer. The rest of the year. The rest of my life.

Last night I had dreamed about her. She stood with her back to me, tanned and young and firm everywhere, her spine like brand-new rope, but when she turned, she was wrinkly and grumpy and resembled Alma.

I shook off the memory and opened a letter addressed to Sverre. It was from the agricultural control office. The annual inspection of the seed potato crops was to be next Friday at nine o'clock.

"The priest told me that Einar had been here, that he came to see my mother," I said putting the letter down.

"When was that?"

"In 1967."

She looked at me for a long time, then she got up, took my head in her hands and looked me in the eye.

"Dear, dear Edvard," she said. "Get your calculator out. That was twenty-four years ago. Something dark and horrible resulted from it. Dear God, look around you! It's summer, you've just inherited a farm and your woman is ovulating."

"My woman?" I smiled crookedly.

"If that's what you want."

I kissed her and she filled my bowl with blackcurrants. "I have two things to say," she said. "You have a habit of brooding and tormenting yourself. If you start searching and *don't* work out what happened, you'll spend the rest of your life looking as though you're in the final of the world chess championship. Your parents are dead, Edvard. The truth is that they're not coming back. No matter what, that is the most important fact, and you know it. You have to accept when something new presents itself to you."

She emptied the thick, pale-yellow cream over the blackcurrants. After a few seconds they popped up and floated on the surface, bursting with red.

Grubbe recognised the smell. He stretched his forelegs and yawned so that his canines were visible.

"Cats shouldn't drink milk," said Hanne when I took the carton of cream and poured a little in a bowl.

"What about cream?" I said.

"Same thing. It gives them stomach ache."

I gave it to him anyway. He lapped up the fatty cream and swept his tail slowly across the ground.

"I need to go to Authuille," I said. "And to Reims."

"Reims? Where's that?"

I told her about Francine Maurel.

"Good God, Edvard. Sverre isn't even in the ground yet. If we're going somewhere, can't we at least go somewhere enjoyable? The fixed point in your life is gone. Why would you burden yourself with even more misery?"

"But she's old. And she's the only one who can tell me about my mother."

She twisted a lock of hair around her middle finger.

"*That* I can understand," she said. "But you could start with a letter, at least."

I had more to say, but it was too much. Because it encompassed a word that had been jotted down three times on a parish newsletter dated 1967. Inheritance. As well as another word, as far from this warm, Norwegian summer as it could be: Ravensbrück.

"Listen," she said, putting her textbook down. "The sheep are all in the mountains. Grubbe can take care of himself. What if we go for a trip after the funeral?"

"Where to?"

"Somewhere warm, naturally. Or to our cabin in Sørlandet. Sun and swimming, it could be worse. Come on. That's what people *do*."

"Come outside with me," I said the following day when the sun was at its peak. "I want to show you something."

We walked out to the plum trees. The unripe fruit was no longer hard but hung densely from the branches with the promise of a sweet autumn. We stared at the green foliage and at the plums that would soon be bursting red and rich. But Hanne was thinking forward and I was thinking back.

She lay down in the grass.

Oh, Hirifjell. With you. Fertile as the ground you lie on.

But I could not remain here.

I took her by the hand and led her up to the flame-birch woods. Led her over to the largest tree, a rough trunk shackled with iron bands that had been tightened to such an extent that not even the metal knew the limits of its tolerance, and we both laid down there and looked up into the canopy.

Hanne arched her body from her neck to her heels so that she could loosen her clothes and pull them off, and soon she was naked in that same arch.

"You too," she said. "Now."

Now and then I had imagined the funeral had already taken place. I had always pictured it in the winter, during the mild period before Christmas. I would stand next to the Star, alone in the car park outside Saksum church, scraping the ice off the windscreen with one of his Karajan cassettes. A snow flurry would blow through a dusty old black suit and I would be the last to leave, standing for a long time with a view of the grave as the snow covered it with a white blanket.

Instead it was a hot summer and I was wearing a new suit we had bought in Lillehammer. Hanne to my left, wearing a grey dress she had bought at the same time. Yngve to my right, he had plenty of suits to choose from.

The flame-birch coffin rested in the nave in a meadow of potato flowers. I had been out before sunrise, spent two hours cutting flowers with a sickle. Every type of potato from his fields followed him to the grave, the colour of the flowers a little different for each variety – even Beate was represented. The polished wood sparkled and cast new reflections with every candle the sexton lit.

It was ten to one. A window was open, and while the trio sonatas drifted through the church I began to listen for steps on the gravel. One thought paid regular visits: what if the improbable happened and Einar appeared in the doorway? Every morning and every evening I had tried to call Agnes Brown, and every time I had hung up without anyone answering. I could send a letter, but my plan was already clear: that was not going to be necessary.

Then I heard a bicycle bell and the crunching of tyres, through the window I heard laboured breathing and something banging against the wall of the church, and when I turned towards the door, there he was.

Noddy.

Wearing a grey Catalina jacket and shiny nylon trousers, he gasped for breath, sweating and mumbling. Yngve and Hanne exchanged glances as Jan Børgum stood in the aisle with twitching eyes. He wiped the sweat off his face, then stuck one hand in his trouser pocket and adjusted his balls.

Music surged from the organ pipes.

More footsteps outside. Jan's mother hurried into the church and grabbed him by the jacket. Jan pulled away and came forwards. I closed my eyes.

"Yngve," I whispered. He leaned towards me. "Can you take care of that?"

"O.K., I'll get rid of him."

"No. Tell his mother he can sit up front."

But they had already sat down, there were plenty of empty places to choose from. A few came for the sake of appearances, including Alma's nephew. Then the bells rang out, the organist turned the page and the old priest strode from the sacristy, his bible filled with loose sheets. Wearing a black kirtle, stately and pale, he walked slowly towards the coffin.

My gaze fell on the altarpiece Einar had restored in 1940. I could not believe that it had been shattered, because every line continued unbroken.

I listened to the priest's sermon, tried to hear if a crack could be found *there*, as Thallaug gradually built it up and told of how Sverre Hirifjell "was a man who God subjected to the most arduous trials", and closed by addressing sin and soul-searching, hate and mercy. His words rang out through the almost empty church.

I began to think about the unkempt grave of my parents, how dismal it looked, covered with brown twigs and moss, and I did not notice that the priest had finished until the silence settled in, a tangible silence.

The coffin was so heavy that it took eight of us to carry it. Hanne, Yngve and the sexton on my side. On the other, Rannveig Landstad, her son and two hired hands from the funeral parlour in Harpefoss.

We stepped outside, walked past Jan's blue bicycle, and when we rounded the corner and stepped out of the shadows, a burning summer sun met us and the beams of light penetrated the flame-birch and made the wood flicker, as though we were carrying a mirage.

We did not have to walk to a grave, but to the funeral parlour's old hearse, a Mercedes they no longer used. Bestefar had to take a detour, to Lillehammer to be cremated.

Then everything began to fall apart, as my senses had when I rubbed at the swastika with Lynol. My legs faltered as we passed through the tarry smell coming from the wall of the sun-baked church, or perhaps it was the smell from Mamma and Pappa's funeral that caught up with me, and I felt myself lift away when I saw their grave up on the hill, so withered and dry, stolen by the Devil. It was as though I were split in two, one part carrying the coffin, the other tumbling through his own life, and I dropped down in front of two chiselled names and a date, knowing I had to answer for them.

5

"SHOULD I BOOK?" SHE SAID. "THEY SAY THERE ARE ONLY two seats left at that price."

I stood holding the telephone receiver and felt torn in every direction. But finally I realised the purpose of having the photograph of my parents by the telephone. It was more than a photograph, it was a question waiting to be asked. Did I want to swim against the current in order to find out how they died?

"No, Hanne. I can't do it. Not yet."

"I'm coming out to yours," she said. "You're not yourself."

"No," I said. "I'll come by your place tomorrow. Don't come here now."

I hung up, went outside and closed the gate. In the old Deutz I drove over to the woodpiles to fill a trailer with wood. Spruce and aspen cut to sixty-centimetre lengths. I drove up to the pimpernel fields, hopped out and sized up the place while the tractor rumbled in idle. A little further up, in the middle of the field, was the stretch of land that offered the best view of the farm buildings. Underfoot, down in the soil, there was life. Sunlight, water and mould, a process as infinite as counting the stars.

It was there that I unloaded the wood and began to build, grabbing more and more armfuls. It had to be high, so high that in the end I had to stand on the trailer to reach the top. By twilight it was done, a level, solid plinth of criss-crossed wood.

Back at the farm, I carefully positioned the front-end loader under the pallet with the coffin and drove slowly, turning straight into the field like it was a plough. Potato grass brushed the sides of the wooden coffin. The wood creaked as I lowered it onto the plinth.

I showered, shaved and changed into black. Drank from the outdoor tap, so cold that my temples ached, and glanced up at the field where the coffin glistened in the evening sun like a gigantic gemstone.

It was eleven o'clock when I picked up a small, sealed cloth bag from the living-room table. After the cremation I had asked to be left alone in the urn room. I had emptied his ashes into the bag, my ice-cold fingers shaking as a cloud formed in the air and I realised that it was *him* I was moving, then I had grabbed another bag of ashes from my coat pocket.

The colours were not identical; the one I had brought with me was darker and a mixture of flakes and powder, but it would do, Bestefar and I were the only ones here in the thin light from the high windows, surrounded by the cold scent of holy stone. I had wondered which book would be big enough. His favourites were Thomas Mann and Günter Grass, but there was another that was even more worn, and I had carried it over to the fireplace, read and burned one page at a time, then gathered the ashes into a metal pail.

It was not my grandfather who was buried by the old priest and Rannveig Landstad. In the grounds of Saksum churchyard, we had placed the ashes of 640 well-read pages of John Steinbeck's *East of Eden*.

The birds were still singing as I walked with Bestefar in my hands. In this state he weighed one kilogram, but I bore the weight of a boulder. When I raised the lid of the coffin, there was a faint glimmer from the balance springs once installed by Einar. Where my grandfather's chest would have been, I placed the suit sewn by Andreas Schiffer of Essen, with the Russian bayonet at his hip. In the remaining light I opened the cloth bag and scattered his ashes inside, dust only, but to me it was as if he took shape again, held my gaze one last time and looked pleased, and he was both alive and dead, dead like the frame in the Leica, alive because I knew he would have liked this. Finally I took the concert tickets and let them fall inside the coffin, dropping them like loose feathers that search for the place on the ground where they will rest.

I lit the pine roots at each corner of the pedestal and they crackled as the flames began to rise towards the cordwood above, which in turn ignited taller flames that licked the sides of the coffin, flame-birch that had been ablaze since the day I was born.

Suddenly I spotted a carved motif in one corner, a squirrel hiding its nose in its tail. The bonfire caught hold and the squirrel disappeared in the fire that crept upwards and framed the coffin. It was enveloped by the fire which now shed light across the potato fields and grew fiercer and fiercer, forcing me to step back so as not to singe my eyebrows.

The wood groaned, tongues of flame rose and flickered above the birch, a golden pattern reflecting another golden pattern until the coffin suddenly turned black with soot, a cloak of mourning that caught fire the next second, then there was a prolonged groan as the fire employed its full might and the flames burned the flames.

And then I made a deal with my grief; I would be someone the dead could rely on.

III

Island of the Storm Petrels

1

I WAS NOT WOKEN BY THE SUN, BUT BY THE SHETLAND Islands. By half four I was out on deck. The white railings were dripping wet; we must have passed through a rainstorm in the night.

A few small fishing boats crossed our path. Otherwise nothing, until a strip of land revealed itself in the sea mist and Lerwick grew out of its contours. Colourless slopes became green fields. Small clusters became houses and cranes at the harbour.

I had one week. Until the farm would make its demands. And Hanne hers. I had driven round to her place with the car packed, and when I arrived, she thought I was surprising her with a trip to Sørlandet. Our parting was muted and confused, with unspoken bitternesses and everything hanging in the air.

"Don't get lost," she said sharply.

A jolt announced the docking of the ferry. I followed the lorries, past a sign that read WELCOME TO LERWICK. REMEMBER TO DRIVE ON THE LEFT in both Norwegian and English. As if we had a colony here.

I moved over to the British side of the road, and only after several kilometres was I confident that I was not going to have a head-on collision. At a vantage point I noticed that the landscape was similar to the mountain reserves back home, only greener. The same heaths. The same sheep. The fields sloped evenly downwards. The only difference was that the landscape was cut in half and plunged to the sea.

The smells and sounds were different from the woods I knew.

Saltwater with added fish guts and the thick smoke of burning coal or peat. The screeching of seabirds, the rumbling of the breakers striking the cliffs at the mouth of the inlet. The North Sea and the Atlantic Ocean on either side, forever throwing themselves at the coast. It was like standing in a besieged fortress.

I breathed in the sea air. A cold, salty wind, fresh yet rotten. I liked and disliked it at the same time; it reminded me of mould making room for something new.

Something was missing. Something I had expected, but I could not figure out what it was. When I had driven a little further I realised, yes, of course: There were no trees, not a single one. Just small thickets, stone houses and pastures. Not even the lean stem of an aspen.

How could a cabinetmaker stand it?

I bought a map and sat in the car. The Shetland Islands resembled a shattered bottle; small reefs and islands were like shards along the length of the coast.

The priest had made out two place names, Unst and Scalloway. The northernmost island and a little town near Lerwick, on the opposite coast. As I plotted my route, it dawned on me that I had spoken proper English for the first time. *A map of Shetland, please. Yes. Thank you.* And it had been easy. Every victory was a victory over Hanne and Bestefar. *Don't get lost?* Yes, I had neglected my English lessons, but I had other teachers to turn to. Joe Strummer and Shane MacGowan had taught me English. A silver Pioneer stereo and the lyrics on the sleeves of L.P.s had taught me English – at least enough to buy a map.

But for Shetland, I realised, I should have learned old Norwegian instead. The map teemed with names from another age, names for travelling in longships, names for mounted warriors. Wick was a bay. Voe was a slightly larger bay. There were skerries everywhere, and Swarta Skerries must mean that the skerries were black, while Out Skerries and Haaf Skerries were reserved for those furthest out to sea.

But this came at a price, particularly for someone embarking on a search. The map had ten or twelve instances of Hamnavoe, even more for Sandwick, and the small islands were either called Inner Holm or Outer Holm, and if they were not called that, they were called Linga.

On Unst, the Norwegian did not seem watered down at all. Bratta. Hamar. Little Hamar. Framigord were the farms closest to the road. Taing of Noustigarth was a tongue in the sea near Nordigard.

I could not work it out. What possible attraction could Einar, the Parisian, have had for this place where everything seemed to be named after a Viking skald? A man who, already in his teenage years, had tired of making kitchen dressers for country estates. He would be well over seventy now. What do you say to someone who has been away from his family for all these years? Was he even going to care that his brother was dead?

All of a sudden I felt like turning back, letting everything be like before. Because the Leica was on the passenger seat and the last exposure was of Bestefar's dead face. I remembered something he once said, it might have been the autumn after I read *Det Hendte*.

"Seed potatoes," he had said. And I noticed in the way that he straightened up from his work, the way that he studied me and in the words he had opened with, that this was something he had been prepared to say only when I was "big enough", and that he had measured me and found that the time was ripe. I do not know what I had done in the preceding minutes to make Bestefar suddenly realise that I was old enough. I thought I had been working as I always did, but maybe I had developed something pragmatic about the way I worked that made Bestefar straighten up like that and say:

"Each potato is the other potato. All the potatoes we put in the ground now are in fact the same plant. Only when we set actual *seed* does new growth form. Those we planted last year, all those we plant next year, are one and the same potato. So called seed potatoes do rot, yes. But the new ones are just growths from the old ones. They are not only family, they *are* one another."

I started smoking that year, and I smoked his tobacco.

I pulled off my anorak and reached for a chocolate bar from the glove compartment. In addition to tinned food and potatoes, there were twenty Gullbrød, twenty Firkløver and ten bags of peanuts as emergency supplies. Tools and spare parts for the car. A box of things from the secretary desk that I had not found any explanation for. I had photocopied the most important documents and made a print of slide 18b, the one that did not fit in with Bestefar's photographs of Germany.

I munched on the chocolate and told myself that it was just a matter of keeping going. Suddenly the wind subsided. Maybe it was a bad omen to break open the emergency supplies on the first day?

A dark-grey bank of clouds was rolling in from the mouth of the inlet. The storms at home announced their arrival well in advance, and always with a muggy messenger, so I imagined the rain would reach Shetland that evening. But the shifting weather blew in as fast as an angry bull. The wind picked up again and fifteen minutes later I had the windscreen wipers on full and was heading for Lerwick, for Agnes Brown's hair salon.

When St Sunniva Hairdresser's closed in 1975, they must have simply let out the last customer, swept the floor and not bothered opening the next day. And so the years passed, until now, when I stood looking through the shop's dusty window. A faded Wella poster was hanging over the entrance, showing the profile of a woman with a wavy hairstyle. On a table was a yellowed copy of the *Shetland Times*, so dry that the pages curled towards the light. It was difficult to see further inside, but I could make out that large, light-blue hairdryers had been abandoned in the middle of the room. Old-fashioned shampoo bottles by the washstands. The place reminded me of Einar's workshop, deserted and yet intact.

I turned and stood under the eaves. The rain ricocheted off the asphalt. It was Friday and people seemed to be shopping for the weekend. They walked quickly, untroubled by the cloudburst. Simply pulled up the hoods of their rain jackets.

All around there were drab stone houses with small gardens. On the way here, walking down King Harald Street, I had passed majestic buildings with spires and round, lead-glass windows that reminded me of the castle on the copy of *Robin Hood* I had back home.

And here, in St Sunniva Street, a tiny hair salon had attracted its customers. A light was on in a window above the salon. I opened an iron gate, walked through a small garden that had not been tended to for some years, and felt a shiver when I saw the brass sign by the door.

AGNES BROWN.

I pressed the doorbell three times. No-one came. A window was ajar. I stepped back into the rain and shouted *Hello*, but there was no response.

Across the street was a clothing shop. A lady with curly red hair was repairing yellow oilskins. The repairs were being made with a Tip Top puncture repair kit, the same I used to mend my inner tubes back home.

"The hairdresser," I said and nodded at the salon.

She put down the tube of glue and scrutinised me. "You don't need a haircut, do you?"

I laughed. "I need to find Agnes Brown," I said.

"She hasn't cut anyone's hair for years. Go to St Magnus Street or King Erik Street. There are good hairdressers there."

"How much do you want for that?" I said, pointing to an oilskin.

"It's not finished yet."

"But when it is."

"Don't know," she said and held it up to the ceiling lamp, perhaps to see how much the tear lowered its value. "Depends on how much you have in your *pung*," she said.

"Hm?" I said, surprised to hear the Norwegian for scrotum.

"You're Norwegian, aren't you?"

"Yes."

"What do you call the thing you carry money in?"

"*Lommebok*," I said.

She repeated it clumsily. "*Loomi-buuk?* We call it a pung."

We made the transaction. "I don't need a haircut," I said. "It's Agnes Brown I'm after."

"She is such a lovely old lady. Apparently won a beauty contest in her younger days. But I haven't seen her in a long time. She keeps to herself."

Her pronunciation was easy to make out. I had thought that the Shetland dialect would have a lot of Scottish influence, but the ring of her words was more like what I heard in the car on the B.B.C.

"Do you know of a Norwegian by the name of Einar Hirifjell? Came here during the war."

She shook her head. "Sorry, no."

"Does anyone else live in the flat above the salon?"

"I believe Agnes has lived alone her entire life," she said, folding up the oilskins. She was in no hurry, smiled as though she wanted to ask me something, but it was too soon to do so there and then. Her calm and open nature carried the promise of a "next time".

I was not used to having no past. Down in the village I was always on guard; here in Shetland, I felt as free as I did in the mountains.

On a whim I pulled out the notebook. "I think this is Agnes' number," I said. "You have a telephone here, I imagine?"

"Yes, of course."

"Could you dial the number? If she answers, then ask if she minds a visitor."

She went into the back room and dialled the number. Poked her head out. "No answer," she said.

"Try again." I bounded across the street to the entrance door in the back garden. Looked up at the open window and waited. No telephone rang in Agnes Brown's flat, and it was small enough that the sound would have reached here.

Back at the gate, I heard a jingling coming from inside the salon. I peeked inside. Next to an open cash register, a grey telephone was ringing.

I got back into the car. The priest had mentioned something about Scalloway, and the old harbour for the Shetland Bus. Ten minutes later I parked the car, put on the oilskins and looked around me. Scalloway was simply a few streets around a small wick.

This is where his life must have taken him. He had crossed the grey, desolate sea to put an unresolved conflict with Bestefar behind him. As I had done too, in a way.

What could a cabinetmaker find to do here in this fishing village, made a headquarters for the war effort by geographical coincidence? I tried to imagine those years. The same landscape, the same weather, though it was only the dark side of everything that mattered.

On the other side of the street was a sign. Royal Mail. Why had I not thought of that before?

The place was full of people but did not look like a post office, more like a second-hand bookstore: rows of shelves with dusty pulp fiction and magazines in orange crates. The only thing qualifying it for the sign outside were two plastic crates: a yellow crate with stamped letters that had not been franked, and a red case filled with newly arrived post. People seemed to be helping themselves to the letters, but soon I made sense of the system; whoever took several letters with them was probably dropping off post to the neighbours they passed on the way.

I waited until the red case had emptied a little and the crowd had thinned before approaching the postmaster, a balding man who was sorting through some comic books.

"Einar Hirifjell," I said quietly. "A Norwegian. Does he live here in Scalloway?"

The postmaster looked up at the ceiling. Seemed to be making a mental calculation. A young man came to pay for some books. Suddenly I had second thoughts. The local gossip would spread, maybe even reach Einar before I did. Our meeting was not meant to take place with a full audience and a brass band. I wanted to observe him from a distance, let the sight of him arrive undisturbed.

"No," the postmaster said. "But some Norwegians stayed behind after the war. And then they took our womenfolk back to Norway. I'll ask Lise," he said and picked up the telephone.

Five minutes later I was in the gentle tentacles of Lise Robertson, a buxom lady wearing a flowery jacket and sensible shoes. She was half Norwegian, and one of Scalloway's guides for the Shetland Bus. Her delivery was like a tightly edited radio broadcast, something she must have been perfecting since 1945, and with a meticulous sense of detail she described how the Norwegian fishing boats had operated a shuttle service between Scalloway and the coast of Norway. Weapons and explosives and saboteurs one way, refugees in return, while low-flying German fighter planes peppered them with machine-gun fire through gaps in the fog.

We hurried around the pier and stopped by a sculpture of a fishing boat riding the waves. Beneath the inscription ALT FOR NORGE were rows of names of Norwegian seamen who died working the Shetland Bus.

"Norwegians have always been popular in Shetland," she said. "We *were* Norwegian. This was Hjaltland, right up until the Scots took over in 1472."

She described the horse-trading that took place when the Danish king married off his daughter and, unable to pay the dowry, he gave away the Shetland Islands instead. These were dark times for the Shetlanders. The Scots did away with Viking law and turned free men into tenant farmers. On the island of Yell, the feudal lord had forced forty men out on a fishing expedition in rough weather. It soon turned into a storm, and thirty-four families lost their fathers and sons. For that reason, the word *Norge* had always had a golden ring to everyone in Shetland, as it was a link to their time as free people.

"Then the war came," she said as we stood on the jetty known as Prince Olav Slipway, "and suddenly they were here again, the Norwegians, in fishing boats from across the sea. Fearless folk, just as we had imagined them. Young and brave. The Germans shot their boats to pieces, but the Norwegians did not yield. Repaired them down at the workshop and set out again the next night."

"Wait a minute," I interrupted. "Did you say they used wooden boats?"

"I said they used fishing boats."

"But they were wooden?"

"Yes."

"And they were repaired here?"

"Right down there," she said and pointed at a dilapidated building near the dock. "Jack met all of them." She tilted her head at a man wearing a boiler suit. He was heading towards a shack, carrying a wooden box which was obviously heavy, as his pace quickened after he had set it down.

We followed him into the workshop, to the sound of chisels and angle grinders, and Lise Robertson managed to persuade Jack to speak to us in his cramped office.

Only one piece of information came to light over the next quarter of an hour. But it was significant.

Einar Hirifjell came here in 1942 and became a first-class boatbuilder.

"That is to say, he wasn't a boatbuilder to start with," Jack said. "But he learned his trade remarkably quickly. As I understand it, he was a cabinetmaker originally. A true miracle. Could repair a splintered hull faster than any of them. They built brilliant hiding places for the weapons. One of them looked like a fish barrel, but inside there was an anti-aircraft gun that could be raised in a flash. But they had to stop in 1943. By then the Germans were using more planes and had sunk every second boat. The crossings came to an end, didn't start up again until the Americans supplied U-boat hunters."

I was listening intently.

"Yes, because of course on a steel ship, there's nothing for a cabinetmaker to do," he explained.

"Ah, I see," I said. "What happened to him then?"

"Hung about unemployed. Did odd jobs. Built fishing-net boxes in exchange for tobacco. Then he disappeared. I heard he had taken on a job for a wealthy man on Unst."

"Unst?"

He scratched his stubble. "Unst," he repeated.

I picked up the photograph Bestefar had taken. "Is this Unst?"

He took a quick look at it and shrugged. "It's on Shetland, at least," he said.

I waited for him to continue.

"Because there are no houses," he said.

The mechanics began to get ready to go home. The lathes and the drill presses stopped one by one, until they were all silent and heavy, surrounded by a whiff of machine oil. Jack glanced at the clock, his eyes revealing that soon he would have to be making a call to explain why he, the foreman, was going to be late for dinner.

I began to walk towards the door, then turned and asked:

"Do you know, by any chance, whether he went to France during the war?"

He shook his head. "The Norwegians got up to a lot back then. 'Ask no questions and you'll be told no lies', they used to say."

Scalloway was already quiet. By the time I left, the Friday evening calm had hoovered up the entire place. The light from the Royal Mail sign was the only indication of life.

There were fewer and fewer cars the further north I drove. When I reached "Great Britain's Most Northerly Fish & Chips" in

116

Brae, I stopped the car, switched off "Brownsville Girl" and walked inside.

While I ate I pictured the farm back home. Deserted for the first time in 150 years. Grubbe had sat on the stairs and would not let me stroke him. He had sensed that I was going to leave.

At the very back of the tool house was a horse cart, the one my great-great-grandparents had arrived in to break fresh ground at Hirifjell. As I left, the buildings shrank into place in the rear-view mirror. The reflection rattled when I drove over the cattle grid, and a moment later I closed the gate and swung out towards the county road. As I crossed the mountain, it was as if I was driving away from the old version of myself, but now, sitting here and eating foreign food, it was like the old me had returned, and I began to question whether I had remembered to lock the gate by the cattle grid.

Full and with the salty taste of deep-fried potatoes in my mouth, I changed cassettes to The Clash and continued north-wards, just caught *Bigga*, the ferry to Yell, before racing on past "Britain's northernmost pub".

Unst greeted me with a light rain, but the yellow oilskins kept me dry. I stood at the very front of *Geira* and felt the shuddering of the ferry's steel hull carry me closer and closer to this rain-laden, desolate location in the sea, as treeless and barren as Yell, ground down by salt wind.

I turned and looked at the car deck. Apparently only heavier vehicles were travelling to Unst. The family cars had stayed behind on Yell. Around the Commodore, there was a Bedford, a flaking Land Rover and a Toyota Hi-Lux with crab pots in the back.

Two bearded men sat in the Hi-Lux. They had been staring in my direction for the entire crossing. Now the driver leaned over to hear what the other was saying. I felt I had set something in motion.

A few more minutes, then we were there. I ought to have gone down to the car, but stayed to watch a man on the dock on Unst, an old man standing on his own with a walking stick.

We drove ashore. The old man climbed into the passenger seat of the Land Rover.

Since Yell, "northernmost" or "most northerly" had been the only distinction for businesses along the way. Most northerly school. Northernmost hotel. Northernmost bus stop. I arrived at "Britain's northernmost grocery shop", where the fluorescent lights were still on. I picked up some fresh food, a few cans of beer, and was amused by the foreign labels and the fact that one was able to buy spirits in a normal shop. A family of three had almost finished their weekend shopping and I strolled past them near the counter.

A freckled man in grey overalls entered the products into the cash till. When he was halfway through I asked about Einar Hirifjell, using the same words I had at the post office.

The assistant reacted strangely, his hand freezing in mid-air as he was about to enter a price on the till. He gave me a sideways look and said, "Isn't he up at Norwick?"

Then it was my turn to react strangely; I dropped my wallet on the floor and fumbled for words. "Norwick," I said quickly. "Where's that?"

"At the end of the road," he said, pointing north.

The family behind me began to unload their shopping trolley. A child was nagging his parents for sweets.

"Do you know Einar?" I said as I took my change. "Does he shop here?"

The assistant did not seem to understand what I had said. I thought to repeat myself, but then went out to the car to study the map. I couldn't find anywhere called Norwick – could I have misunderstood his pronunciation? I waited until the family had come out and went inside clutching the map.

But the shop assistant was no longer at the till. He had gone into an office behind the dairy counter, and now stood with his back to me, speaking into a telephone.

I stepped outside. The island was saturated and quiet, numbed by the evening.

I had overheard something and it made my plan seem brash and egotistical. But perhaps I had heard wrong? Perhaps the swooshing of the fan above the dairy counter had jumbled his words. I distinctly heard "the Norwegian". What had he said next? – "waiting", or "wanted"? Or had he said what I had at first discounted as improbable: "Someone's come at last"?

He said it as a warning, the kind that makes a turtle retract into its shell. Perhaps Einar did not want visitors. Perhaps the man who had made a name for himself in Paris with a jack plane and a surface gauge had become a grumpy old man who didn't answer when someone came to call. Many years had passed since my tenth birthday, and his life may have taken a turn.

No, I told myself. For a man who fells birch trees with a bow saw, builds a coffin from its wood and sends it to his brother, life moves very slowly, or does not move at all.

I drove around without finding Norwick, but then I thought I recognised a view similar to that on the photograph Beste-far had taken. I followed a small road inland. What had made a middle-aged farmer with no interest in photography take *one* photograph of an insignificant stretch of coastline?

The road led towards the southernmost tip of the island. I climbed out, pulled on my anorak and tried to get the terrain to fall into place. The drizzle came and went as I continued my search for a photograph that had already been taken.

A dog was barking behind a ridge. Soon I heard it again, closer. A bedraggled, grey-spotted pointer trotted past with its tongue hanging out, its ears pressed back and wild eyes, and then it was gone. Down the side of the ridge a breathless, scrawny woman gave chase. She hurried towards me, opened her jacket and closed it again. Seemed even more restless than the mutt.

"It ran off that way," I said, pointing.

She scratched herself roughly with her left hand. Her skin was scabby and inflamed.

"Is there a Norwegian living in the area?" I said.

"Not that I know of."

119

She followed the dog, came back a moment later and shook her head. Caught her breath and pulled out a pack of Salem menthols. After a few drags she pinched the ember, put the cigarette back in the pack and scratched the back of her hand again. "Do you know where this is?" I said and handed her the photograph.

"Haaf Gruney" she said nodding, and pointed at an island that was barely visible in the background of the picture.

The dog came back wagging its tail and sprayed water against my leg. She attached its lead.

"Are you superstitious?" she said, forcing the pointer to heel.

"Not really."

"People used to believe that you could see the Devil rowing across the strait from Haaf Gruney. He would arrive at night with a coffin sticking out of his boat."

I stared at her.

"Coffins?" I said. "Did someone live out there who made coffins?"

She didn't answer, pulled out the cigarette and was about to relight it when the pointer raced off again.

"Where's Norwick?" I shouted after her.

She turned. "Where the coffins arrive."

The map was getting wrinkled. Haaf Gruney was between Unst and the neighbouring island of Fetlar, and it wasn't far. Gusts of wind tore at the paper. I put the Leica over my shoulder and started walking.

When I reached the waterline it was as if geomagnetism was guiding me forward, correcting me like the needle of a compass. The landscape fell into place; the ridge to my right had the correct height, the inlet to the left had the curve I was looking for, and soon I was inside the photograph.

Bestefar had stood *here*.

Instinctively I looked at the ground, as though his footprints were still visible. It *was* here, I even recognised this stretch by the focal length of his Rollei camera.

But everything was even less comprehensible. Because around me there was not a single house. Just the road, the sea, and a ramshackle boathouse built of grey stone.

I fetched Bestefar's binoculars from the boot. The German optics removed all doubt: Haaf Gruney was uninhabited. Not so much as a drystone wall to be seen.

My last hope was Norwick, but it was too late in the day to be knocking on doors asking about Einar. I kept driving around, to discover the island's perimeter and to find a place where I could sleep the night in the car.

Then I found Norwick. There was no sign, but as a Norwegian I should not need one. Because when I arrived at the north end of the island, I was met by the same sight that had made some Norseman give a name to that most northerly wick.

The broad mouth of a fjord into which the sea roared, six or seven houses on a hillside. A small cemetery at the tip of a headland.

Some kind of animalistic trembling took hold of me, as well as the acute certainty that something would soon be settled. He was here, somewhere nearby. I opened the iron gate, stepped amongst the gravestones, and walked towards the one that stood out.

A coffin made of flame-birch crossed the North Sea.

A gravestone made of greyish-blue Saksum granite arrived in exchange.

2

JULY 1986. I REMEMBER THE MONTH WELL. BESTEFAR had returned from town with the Star weighing heavy on its rear axle, like in winter when he would drive with sandbags in the boot to make the studded tyres grip. There had been a gravestone in the back, though his explanation was that a rear spring had snapped and would have to be repaired in Lillehammer. This would fit in well with the annual meeting for the Association of Sheep and Goat Breeders, which had been moved forward a little.

That year Bestefar must have called me from here, perhaps using the phone box down by the ferry. But he hadn't told me that he had buried his brother. Later we cancelled the P.O. box in town and started using the postbox along the county road.

I sat on a bench in the cemetery and spent a long time studying the gravestones. Weather-worn, covered with moss. Each one a memorial stone to everything I had hoped to discover. The truth about my parents. The four days of my disappearance had a gravestone of their own, contorted and bleak. Just beyond, a rusted cross. To commemorate what the inheritance might have been. At the edge of my field of vision, an arched stone, beneath which lay the answer to why, why, *why* Einar had to be concealed from me. At my feet, a small white stone with weathered dates, a monument to the mystery of how Mamma and Pappa had met.

And largest of all in the middle of the cemetery, a towering monument above a grave that opened up as I sat there, the hole inside me.

I went to Einar's gravestone and crouched down. Norwick was a place so weather-beaten that even the flowers at the graves had to be lashed to small metal stakes. Beside Einar's flapped a frayed piece of yellow string. A little way away, in the direction of the wind, orange tulips were scattered in the grass. I had thought they belonged to another grave, but now I saw the same yellow string around some of the stems.

Very recently, someone had left flowers at Einar's grave.

I retied the tulips and stepped away. The wind still tugged at the flowers, but now it was as if the intention was to scatter the petals and spread the seeds.

The blue flames of the Primus swooshed to life. I set up camp with a view of slide 18b and now sat leaning against the rear wheel, eating pea soup. Apart from the boathouse down by the shore, there was nothing to see. A few sheep bleated on the hill-side, but otherwise the landscape was lifeless. It was approaching eleven, I was exhausted and getting cold. The ferries would still have been running, I could have hopped in the car, checked in to a cheap hotel in Lerwick and gone to Agnes Brown's the next morning.

My gaze came to rest on Haaf Gruney out in the strait. The fact that Einar had rowed coffins at dusk might correspond with what I knew of him, but he couldn't possibly have lived there. Still, superstition had attached itself to the flat and peaceful island.

The fog began to seep in. I grabbed the Leica, walked down towards the rocky shore and found an overgrown path that led to the old boathouse. I positioned myself on the leeward side. It was low and constructed of grey stone with a rusty, corrugated-iron roof. The wall jutted into the sea a few metres so that boats could be rowed directly inside.

A few flat stones above the surface of the water led to the end of the boathouse. I followed them, holding on to a rope that ran along the roof, and down around the corner. The sea splashed at my shoes. A large wooden gate, splintered and chewed up by the sea and the weather, barricaded the entrance. In its centre, in flaking white paint, the traces of a large X, a saltire.

And a Norwegian Mustad padlock was hanging from the sliding bolt.

I ran up to the car and searched through my luggage for a torch and the key ring I had taken from the secretary desk. I looked up the road, listening out for people. Raced back down. Grabbed the rope and stepped towards the gate with the water sloshing around my shoes.

The key slid in. A small click and the metal loop popped open. Cradling the torch in my arm, I lifted the sliding bolt, opened the gate and slipped inside.

The evening light revealed an old rowing boat pitching in the water, in this confined continuation of the sea. The back of the boathouse was on dry land, like a cave, barely a metre to the roof, and I glimpsed something white and rectangular there. I closed the gate so that no-one could see me and switched on the torch.

The beam of light shone on a white coffin that had been smashed to pieces.

I do not know if I screamed or if I *heard* a scream. My heart was pounding like a rabbit sensing it was about to be slaughtered. The torchlight was fixed on the shattered coffin and I dared not move it, in case there were worse sights lurking in the boathouse.

An uneasiness passed through me, and a fear that the coffin might contain a rotting corpse. Not a body with bones and worm-eaten skin, but one I would never be able to bury. A truth I would not be able to handle.

I crept slowly past the boat and got down on one knee. The head of the coffin was crushed. The side panels remained intact, but the joints had slipped apart, leaving the coffin twisted and deformed. I pushed the lid aside.

Fishing nets and offcuts.

Perhaps it was not that disrespectful to store them in a coffin, especially if it was well made. Like the difference between hallowed ground and cultivated ground.

In style it resembled the coffin he had made for Bestefar, but it was barer and simpler, with only a thin frieze, like a braided rope, running along the lid. I shifted the beam towards the boat. Black as coal, simply made and large, almost like a lifeboat. Plenty of room for transporting a coffin.

The torchlight moved across the stone walls. A faded jacket. Coils of ropes. Smeared oil cans, rusted hand tools. Oars.

Amidst the incessant thumping of the sea against the rocks, I could hear the engine of a car changing gears. I switched off the torch, went out and glanced up at the road. A pair of headlamps was visible in the fog. A car idled by the Commodore for a moment before speeding off.

A little later I slid open the entrance gate to let more light into the boathouse. Everything was wet, the sea was sloshing against the stones. The boat was rocking gently back and forth, moored with a slimy length of rope, shaggy and overgrown with green algae.

I noticed a name on the front of the hull. At first I thought it was "ATNA", until I saw the outline of a fifth letter. PATNA. It must have belonged to Einar. In which case it had been here for some years without sustaining damage. The woodwork was swollen together and tight, and covered with barnacles.

Across the strait I could make out the contours of Haaf Gruney. The boathouse was situated at the shortest distance to the island. Maybe an uninhabited island was not the same thing as an *empty* island?

The boat creaked, it sat low in the water and turned sluggishly. I headed for a crag silhouetted against the sky, plotted a course that kept Haaf Gruney to my back, and rowed. Never had I rowed something more cumbersome, but it was solid as a mountain, and perhaps that was how coastal boats were meant to be. A little reluctant.

125

More of Unst was revealed. Light shone from a cluster of houses at the tip of a promontory, and as I got further out, the lights from more houses gradually became visible.

I took off my shoes and sat barefoot so that I could feel if any water began to trickle in. But it was not necessary. Even though the boat was crude and bulky, I sensed that Einar had built it. I remembered the old defiance I had seen in the flame-birch woods, the perpetual competition between two brothers, my inclination to take Einar's side whenever I was upset with Bestefar.

The sound of a boat engine drifted across the water. I raised the oars and looked around. Either the boat had no lanterns, or it was hidden behind Haaf Gruney. The sound bounced off the surface of the water and the surrounding islands, and for a moment it seemed to be heading straight towards me, then it veered off, grew fainter and finally I no longer heard it.

The island was close now. I adjusted my line, rowed quickly through the night sea.

Haaf Gruney loomed larger. I heard the sound of water lapping at the rocky shore. The moon broke through. From a distance the island appeared flat, but the approaching shore was four to five metres tall, lined with craggy rocks. I would have to row to a better landing spot.

Soon I found a shoal, tested for the bottom with the oar, rolled my trousers up and jumped in. But the boat did not tip as I expected, and I struggled to stay on my feet.

I tied the boat, pulled off my anorak and staggered onto the island. I found a pool of water and drank, then sank down onto the grass.

Fatigue began to drift in.

That morning I had come ashore in Lerwick. Now I was sitting out here. If things continued like this, I thought, I could be on the South Pole the day after tomorrow. I pulled a wet chocolate bar from my anorak. *Now* was the time to break out the emergency rations.

The lamps on Unst cast long, yellow shimmers across the surface of the sea. As I ate the last bit of chocolate I saw a new light, fainter, it did not reach the sea and must belong to a house further inland.

Haaf Gruney was covered in tall, bristly grass. I walked inland, climbed up on a large rock.

A view of night and wind.

I took off my trousers and wrung them out, put them on again, stiff and clinging, and searched for driftwood for a campfire. The cold began to take hold. I tramped about, looking for a place that could afford the best shelter.

If there *was* shelter on Haaf Gruney.

I walked the whole length of the island and found nothing. Nothing but pebbles, puddles, the sky above me, the surrounding sea. I turned back and walked to the southern end, following the curve of a slope, and expected it to end in an abrupt cliff above the sea.

But instead I found myself looking down on the roof of a house. The straight lines broke with the scoured terrain around me that seemed to have been untouched for ten thousand years.

Two, no, three small stone buildings. A small boathouse by the shore.

All invisible from Unst.

I staggered down the slope between the walls of the buildings, and came to the front of the largest structure. I fumbled in my anorak for the key ring tangled up in a mess of chocolate wrappers and soggy pound notes.

3

WAS IT EXHAUSTION OR A FOREBODING THAT MADE ME knock on the door of a house left by a dead man and shout *Hello*? Who can say? Perhaps it was simply meant as a greeting to Einar's ghost, a ghost unable to voice its reply, but there all the same. I stepped inside quietly, my movements echoed in the empty hall, and it was as though I could feel his presence in the night.

I made it, Einar, I thought. *I do not know if you actually wanted me to come. But I believe you did. You arrived at Hirifjell when I was ten, and I did not know about you then. I arrived too late, but I'm here now. So show yourself in any way you can.*

Shutters were fixed over the windows. I groped in the darkness, extended my arms and felt my way forwards along the walls. Aware of the faint smell of soot, I ran my hands along the brickwork of a fireplace. There, on the shelf, a small box. I shook it. The same reassuring sound as in Norway. Matches.

The flame revealed a table and a couch. A bookshelf under a window, otherwise nothing. I gazed around in the light of the dying match. Struck a new flame and found a yellow candle, but nothing to get a fire going in the fireplace with, nor was there any bed or blankets. I was cold and searched the room for anything soft and dry, ended up tearing down a curtain and wrapping myself in it.

Just as I was about to fall asleep, something jerked me awake.

The boat. It was still moored where I had made my foray. I had no idea whether it was high tide or low tide now, or whether it might work itself loose. Quickly I pulled on my wet shoes, wet trousers and anorak and ran outside.

It was rocking beyond the stones and seemed to be welcoming me back. On Unst, I saw the lights of a solitary car passing.

With my last strength I rowed around the island. Barely avoided a jagged reef just in front of the dock. I tried to open the boathouse but could not find the right key, and ended up mooring the boat to a couple of rotting poles.

That night my dreams spun in circles.

I stood in a large room with a woman in a dress. Light from high windows cast shadows across the floor. We stood quite still. As though we were waiting for music for a dance.

We were the same age, yet she was grown up and I was not. We embraced but could not feel each other's touch, it was as though she was air to me and I was air to her.

There was something that did not feel right.

And then her figure began to fade. The dress still held the shape of a person, but then the material lost its form, the thin fabric collapsed. I caught her waist with my arm, and stood with a dead woman's dress, alone.

I woke and wondered whether a dance was about to begin. Or would never end.

Outside I heard the breakers pounding Haaf Gruney. I fell asleep again.

A hazy light filtered through the cracks of the shutters. My clothes were lying in a wet heap on the floor. The dream still trembled inside me, like a phantom etched on my retina, gradually expelled by daylight.

The world outside was grey. The sun was attempting to break through. Some overgrown flagstones led down to the boathouse, and the rowing boat was where I had moored it. The wind was calm, but the sea was still broken up against the reef beyond the boathouse.

I loosened the wing nuts that fastened the shutters from the inside, went outside and lifted them off. I could see the light enter the house, like when I had reconnected the electricity in the carpentry workshop.

Then I walked over to the front door, grabbed the doorknob and told myself that I was Einar Hirifjell.

A creak from a sluggish door. Small hints of life. Worn paint at the bottom of the wall where he had kicked off his shoes.

Some of his daily rhythm became evident. Waking in a small bedroom beyond the kitchen, where an old spring mattress now lay bare on a single bed made of rough wood. A dented washbasin and a green metal ewer. A towel and a dried-up piece of soap.

The coffee pot upside down. Breakfast alone. A solitary stool by the kitchen window at daybreak. A view of the nesting cliff on the neighbouring island.

The furniture he had made, I could tell by the joints. But they were as simple as a carpenter's bench. Enthusiasm for his own work, elegance, did not have a place in this house.

He used peat for heating. In a box by the kitchen stove I found some dry, black clumps. The floor was worn in an arc between the kitchen and the living room, where he must have settled on a small couch and placed his cup on a low table.

A Kurér radio on the windowsill. Smoked pipes in a brown latticed bowl. Nothing appeared to have been touched since he died.

I sat in the dim light by the hall. The wind blew in through the open door.

Einar Hirifjell. Between stone and sea. Rain and wind. A harsh sky above a haggard man.

A gust of wind swept through my hair. After the funeral Bestefar must have changed the locks here. But why did he take the photograph from Unst and not from out here, and why had the place not been sold?

I was cold. In the hall I found a faded, greyish-green outer jacket, a pair of oil-stained work trousers and some mouldering, pale-yellow Dunlop wellies. *Apart from that he was rather worn and frail,* the priest had said. *Wore ugly yellow wellies.*

I strolled around the small courtyard. It was like finding an abandoned summer cottage that I had discovered belonged to the family. Half mine, half another's.

I have to find the key to the boathouse, I told myself, to get the rowing boat inside. To avoid being discovered. Avoid speaking to anyone. Avoid finding out that the island in fact belongs to someone else.

Beneath the gutters were wooden barrels filled with rainwater and I bent over and drank, spat out strands of algae and drank some more. Then I heard the motor of a boat.

I dropped to the ground, hoping it was a fishing boat passing on the other side of the island. But the droning of the motor grew louder, and soon I glimpsed the bow of a boat rounding the reef. It slowed up in front of *Patna* and remained rocking in the sea.

She stood with one knee on the thwart and one hand on the wheel. A small, scratched-up wooden boat with an old forty-horsepower engine at the back. She was my age and wore a quilted vest. When she caught sight of *Patna*, she brushed the hair from her forehead and looked in my direction.

I stood up and she gave a start. But she did not acknowledge me, just stood there watching, as though standing in front of a freshly painted house having wondered how it would turn out. Then she opened the throttle to take the boat in a wide arc, returned with the sun at her back, moored at the boathouse and climbed ashore.

Mist rose from the surrounding grass. She was not very tall, quite sturdy, and with curly dark-brown hair. Not the type to stand out in a crowd, but still, there was something about her when she came up from the glittering sea. She did not smile as she walked towards me, nor when she stopped a couple of metres away and asked what I was doing here.

"I came yesterday evening," I said.

"Yes, but what are you *doing* here?"

Rolling Rs. Long, low O. Different from the Shetland dialect, she was speaking Scots. Her voice did not fit with her face. She looked defenceless, but had the tone of a bank director.

"Looking around."

"Is that so?" She took a step closer. "Is midnight considered evening in Norway?"

Had it been wrong to say evening? I searched for the right words. "What do you mean?"

"It was past midnight when you rowed across. I was out for a walk and saw you when you were halfway."

"How do you know I'm Norwegian," I said. "Is it the way I speak?"

"Well," she said, walking past me. "You sound like a foreign doctor."

"And that's how you know?"

"No." She looked past me at the buildings. "I know because there's a car with Norwegian number plates near the boathouse on Unst."

Her eyes were brown and steady, as though made to measure and not to admire. She had a habit of narrowing them slightly before she spoke. When I told her about Einar, that I was a relative of his, something dreamlike passed through her, disappearing the next moment.

"You could have waited until daylight," she said. "Got someone to take you across in a proper boat."

I shrugged.

"So why did you come out last night?"

"To let the river run its course."

She chuckled, but in a patronising way, as though mine had been a clumsy but almost respectable response; stupid and pompous perhaps, but hardly something a foreign doctor would have said.

"And you," I said. "Do you come here often?"

She shrugged and strolled towards the buildings without checking to see if I was following.

Now what, I thought. Was I supposed to sit down and pretend I had something to do?

"I come here on the odd occasion," she said when I caught up with her again. "Wander around with a basket, see if I find anything attractive washed up on the beach."

"Do you? Find anything?"

"Now and again," she said. "But you won't fit in my basket."

Her trousers hugged her broad hips. She had round thighs and small breasts. Her sensual face and her arrogance created an undertow that forced me to tag along after her, and the second I realised it, I was annoyed with myself.

"Who owns the island?" I said when we reached the buildings. She frowned and stared at the wrought-iron key in the door, the rest of the bunch swaying back and forth in the wind.

"I mean, now," I said. "Since he died."

"The Winterfinch family owns this island," she said. "Always have done."

"Do they live on Unst?"

"In Edinburgh. They're sometimes here during the summer."

"Do you know them?"

"Everyone knows the Winterfinch family," she said absently and peered into the hallway. Then she stepped back and pointed at the roof.

"Do you know why it's so sturdy?" she said.

I had not noticed it before. The roof was covered with thick slabs of stone which had wire netting stretched over them.

"The netting protects against the sea spray," she said. "So the roof tiles don't get torn off. I wonder what it's like here during a storm. The tallest waves probably reach all the way to the windows."

She was so close to me that I could read what was imprinted on her buttons. *Cordings*. I had never heard of Cordings but imagined they were priced in the Leica range. I searched for something that could place her, explain her. It was as though she was older than me. Not in years, but from another era.

At last I found a suitable term: she was *ladylike*. Her calm, steady movements, the way she had elegantly climbed out of the boat, something endearing hidden behind the controlled expression on her face.

She went over to one of the outbuildings and tugged on one of the padlocks. "How is it that you have keys?"

"I found them back home," I said. "I think my grandfather changed the locks when he was here to bury him."

"I don't think they've been out here for years," she said. "The Winterfinch family, that is."

"Did my great-uncle lease the island from them?"

"I think so, somehow. Why are you wearing his clothes?"

She was so confrontational. Parrying with questions. She seemed like someone who had to fight to get ahead, and one of her weapons was to make others feel simple. "Mine got wet," I said. "This was all I found."

"Not so strange. Those were the only clothes he wore."

"So you knew him?" I blurted. "You knew Einar?"

She repeated his name. Pronounced it *Aainarr*. "I saw him now and again when I was younger. An *unken* body."

She realised I did not understand.

"*Unken*. Eccentric. Recluse. One you do not visit."

"I met a woman," I said. "The superstitious locals apparently believed that the Devil lived here. That he rowed across with a coffin in his boat whenever someone was going to die."

"Not the Devil. *Death*."

"Death?"

"Yes. Because of the coffins. *Aainarr* made coffins. Rowed them over to Unst, and the funeral parlour in Lerwick transported them from there. The story came about because to begin with he had such a small boat that the coffins stuck out at an angle. Of course people around here realised that it was not the same coffin each time, but it was worse with the tourists. Then he got that one." She looked in the direction of *Patna*. "Plenty of room for a coffin."

"Did he have any friends?" I said. "Other than at the funeral parlour?"

"I really have no idea." She strolled towards the rowing boat but stopped at a distance, as though something was keeping her from getting close.

"Probably a dory originally," she said.

"A dory?"

"Yes. See how sturdy it is? They have to be, so as not to get crushed against the side of the ship. Good for whale hunting as well. Typical for a Shetland boat of that size. There were hundreds of them left here on Unst after the old herring catch ended."

Her gaze did not leave the boat.

"Imagine that he died under it," she said.

That stung. Not just what she said, but that I had not given it any thought, how he died. I had pictured him dying in his sleep like Bestefar, one day the candle was simply extinguished.

"Didn't you know?"

I shook my head.

"He *was* family, wasn't he?"

"Of course he was," I said. "But he and my grandfather didn't speak after the war."

"Why was that?"

"They—" I cut myself off. "Tell me how he died," I said.

She pulled the puffa vest more tightly around her. "A fisherman passed by here some five years ago. Saw that the boat had been dragged ashore. He assumed that *Aainarr* was repairing it. But when the fisherman returned with his catch, the boat had dropped and the steel cables of a winch were twisting in the wind. The boat had toppled over and fallen on top of him."

My stomach tensed. I could almost hear the creaking. Wood on stone. Wood on bone. Nobody else on the island, only the wind. A monotone funeral march for Einar Hirifjell.

All of a sudden I could see him. A young man who felt ill at ease in Hirifjell. The dependable employee at a world-class cabinetmaker's, who was sent to Africa to secure the best, *the very best* bubinga blanks.

"So he died alone?" I said.

I said it because I had to say something. A fondness had flared up within me, but it had nowhere to go. Like a stunned bird trapped in a house.

"*Aainarr* was always alone," she said. "That much I can tell you. Someone called for the priest, got the boat back in the water."

I glanced at the stones, looking for specks of blood. Absurd, since I knew they would have been washed away by the first rain shower. Just below the waterline I thought I glimpsed some brass nails. The repairs had been completed. The final blow of the master cabinetmaker's hammer.

"Why was he considered . . . what was it you called him?"

"*Unken?*"

"Yes, that."

She set off towards her boat. When we were some distance from *Patna*, she said in a gentler voice: "I don't like to speak ill of people. But he's dead now, and since you're asking—"

"Yes?"

"There was a story about him. They say he killed a family in France."

"He did *what*?! Why?"

"Greed of a kind. They said that it was over something that was worth a fortune."

A man dies. He leaves behind tools and books and clothes. But he also leaves clues.

In a wardrobe I found a box of shotgun cartridges. To shoot the occasional seabird, presumably. The weapon was not to be found. On the bookshelf were some yellowed issues of *Aftenposten* from the late seventies. The poems of Olav H. Hauge. Some old novels in French. One of them apparently read so many times that the cover was worn to tatters: *Lord Jim* by Joseph Conrad.

Below the novels was a row of dictionaries from French into

just about every European language. Polish, Hungarian, German, Czech, Romanian. All printed in the years immediately after the war. I pulled out the French to Russian *Dictionnaire Larousse*. Well used, threadbare cover. But no forgotten slips of paper, no notes in the margins.

I looked out of the window. The wake from her boat had subsided. Even though I liked the company, I could not wait to get rid of her. She had seemed so rooted here, made little comments that clashed with my right to be here. Because I was the one with the keys. An emissary sent to Haaf Gruney by the dead.

The radio sputtered to life when I turned the knob. Above the crackling, a voice began in Norwegian:

And now a weather warning for the Lindesnes coast—

The batteries lasted a few seconds before the voice of the meteorologist disappeared and left a whistling sound which died too. The tuning knob was stuck. A radio fixed for ever on N.R.K. long wave.

I was faint with hunger. In the larder I found some rusted tins of food. Jenkins Cod Cakes. Perhaps what he had been planning for dinner before the boat tipped over.

Her words were still ringing in my ears. *Greed of a kind.* It must have had something to do with the inheritance.

I fired up the kitchen stove. The peat burned sluggishly, but I began to get warm. Boiled some water and added the fishcakes, and sat staring out of the window.

I think Einar knew what happened to your parents, he just wouldn't talk about it, the old priest had said.

I might be sitting in the kitchen of the man who murdered my parents. But I knew how local gossip travelled, always spiralling to the worst conclusion. Lurking somewhere nearby were the shadows of what had actually happened.

I went outside. Even though the buildings were not visible from Unst, and on Fetlar there was nothing living apart from seabirds in the cliff, it still felt as though I was being watched by someone, either at a distance or quite the opposite. Like an eye within me, without me knowing who was looking through it.

We had not offered our names, the strange girl and I. But she told me that she had grown up on Unst, studied on the mainland and spent the summer holidays here.

"I'll drop by tomorrow," she had said, "and we'll see what happens. In the meantime I'll keep quiet about you poking around out here."

She started the outboard motor, but immediately switched it off. "Be careful," she said and nodded at Einar's boat. "Out here, a storm can arrive in minutes. You've heard what happened to the two girls in 1745?"

I shook my head.

"They rowed over from there," she said, pointing to the neighbouring island. "It's called Uyea. Then the wind got up, they drifted out to sea and ended up in Norway."

She told me that Haaf Gruney had been grazing land for cows at the time. It was on the return trip that the girls were carried out to sea, and they survived because they had milk to subsist on. The storm carried them across the North Sea and they stumbled ashore in Karmøy, where they ended up getting married and having children.

"That's why," she said, "when you have to cross the strait from Haaf Gruney, look to both sides, as if you're going to run across a motorway. Jump in the boat and row as quickly as you can. Otherwise you'll end up married in Karmøy. And that would be a shame."

Then she accelerated suddenly, got the boat to plane and turned towards Unst.

My thoughts kept turning to her as I continued my search. *But you won't fit in my basket.* Said with a playfulness that echoed inside me.

The power of attraction has many forms. In her it took the form of self-confidence, the way she appeared like a righteous messenger with hundreds of longships at her back.

Incomprehensibly, she had awoken something inside me. A desire to show my true self, that I could do more than stand there fumbling for words.

I unlocked one of the outbuildings. Crude tools for the soil. Shovel and pitchfork. Crowbar and sledgehammer. Steel wire, misshapen from being stretched and coiled many times. The pitchfork was missing a tine, the potato-grubber had a new handle. He must have grown a few vegetables to keep the scurvy at bay.

Along the wall was a huge stack of peat, black and greasy and cut into brick-sized blocks. Apart from the radio, there was nothing on the island that had been invented after 1900.

The exception was a strange contraption in the corner. The motor of a Norton motorcycle was bolted to a pallet. A cracked drive belt was stretched between two wheels and hooked up to a dynamo: a homemade generator. The wire ran along the ground and out through the wall. I followed it to the other outbuilding and unlocked the door.

His workshop, fitted out just like the one at Hirifjell. The same positioning of the lathe, the wall with the hand tools set out in the same way, with the same pencilled outlines. The same codes on the bottles of linseed oil. Frayed brushes in jam jars of turpentine. Screws and nails in round tobacco tins. He had smoked Dunhill Early Morning Pipe and nothing else.

Still, the workshop was different from the one back home. Everything was in the same place, but the layout was more *precise*. The wood chisels hung so straight and even that they could have been the marks on a ruler. The templates for the woodcutting machine were stacked like expensive porcelain plates in the service cabinets of the rich. No spontaneity, no small wooden figures, none of the playfulness I had seen in his sketchbooks back home, where he did not accept any design for what it was without twisting it into a series of variations.

I swept the dust from the carpenter's bench. I too had had regrets and torments, and I had kept them at bay through hard work. This simple, spartan location had not been home to a man who gorged himself on profits. Rather, it was the altar of someone doing penance.

The electricity from the generator was used to power the work lamp, the lathe and other tools for which hand power would not suffice. Nothing else. Not even the house had electricity.

The shelves along the side wall held his timber. Oak, pine and a number of varieties I did not recognise. A crate of off-cuts from a dark species of wood, the shattered stock of a gun. On the floor were some light, almost gleaming planks of wood.

I reached for a hand planer, fastened a piece of wood to the bench and made a few strokes. Watched the wood shavings curl away. I wet my thumb, rubbed the wood and saw the grain appear.

Flame-birch. Hirifjell birch. The veined patterns flared up, but not all at once; it was a second or so before the moisture penetrated, as though the flames were chasing my hand.

I tried to determine how he worked. How he thought.

I managed to acquaint myself to some extent with the first, but what he *thought* was a gaping black hole.

This had been his life, waking up every morning to the sea and weather that changed in an instant. A life with Dunhill Early Morning Pipe and a mystery.

Again I attempted to *see* him, as though I was holding the Leica and searching for the one detail that might tell me who he was. This rocky speck in the sea, tormented daily by rain and storm. This snug workshop with its yellow lamplight, warmed by a small iron stove.

Einar Hirifjell alone here.

I continued to explore. Looked under a rag rug, opened the cellar hatch in the kitchen, looked behind cupboards, searched for loose boards – and found nothing. Not until I went out to the carpentry workshop again, looked behind a few tins of lacquer and varnish and discovered a pile of letters. Addressed to me.

Every birthday and every Christmas I had received a letter from Einar. And, just as regularly, Bestefar had returned them.

The fine handwriting I recognised from his Parisian sketch-book had become drawn out and rougher, but just as upright; it was as though every sentence had been made with a scalpel and a ruler, cutting off the top and bottom of the rows of letters. Once or twice he had spelled my name the French way.

Merry Christmas and a Happy New Year, Edouard. I hope you like the present. Best wishes for 1976. Greetings from Einar.

The present? I had never received any present.

The letters were more or less the same each year. Neutral, no reference to anything we might have in common.

There was also a parcel wrapped in tattered, shiny paper, with a checked pattern showing through. A chessboard. Hinged in the middle, with room for the pieces inside. The white squares in light flame-birch, the black squares in dark walnut. The transitions were sharp and precise, the woodwork gleamed with polished wax. Along its length a row of letters had been carved, so precise that it could have been a line in a printed book.

To Edvard from Einar on the day of your confirmation.

On the parcel there was an address slip with Norwegian stamps. It had been returned from Saksum on April 12, 1982.

A brother's name in a brother's handwriting.

Einar Hirifjell

Haaf Gruney, Shetland.

Inside the chessboard, amongst the pieces, there were three newspaper clippings. As I took them out, a rectangular piece of cardboard fell to the floor.

A French identity card from 1943, issued by the occupying power. A swastika stamped over a stapled passport photograph. Einar as I remembered him from the picture in the envelope back home. He had a peculiar hairstyle, a side parting with flat curls hanging over his forehead.

But the name on the identity card was not Einar Hirifjell. It was Oscar Ribaut, born in Paris, with his profession listed as *ébéniste* – cabinetmaker.

*

Ribaut. I had seen that name once before, on the tracing of the message Einar wrote to my mother in the priest's office. Next to the name Isabelle Daireaux.

I studied the picture again. It *was* Einar, but the question of his name soon gave way to another that was even more pressing. Why had Einar, living out here on Haaf Gruney, saved a clipping from a local French newspaper, *Le Courrier Picard*, dated September 1971. In the days following Mamma and Pappa's death.

Taking up three columns, the headline read "TOURISTES DÉCÉDÉS À AUTHUILLE. UN ENFANT DISPARU." Tourists dead in Authuille. A child missing.

I was reading the news about Mamma and Pappa's death at the time it happened, as interpreted by a local journalist and without the benefit of hindsight found in the summary in *Det Hendte 1971*. This was from Friday's paper, barely twenty-four hours after they died:

> A child of three was reported missing near Authuille yesterday morning. The parents, Norwegian tourists, were found dead in a wooded area north of the village. The couple had drowned in one of the many ponds of the Ancre. Judging by their injuries, they stepped on an unexploded gas shell and fell into the water unconscious. The child, a boy, is believed to have become lost either before or after the incident, and extensive search parties spent yesterday looking for him.
>
> The accident is likely to have occurred at night or early in the morning. The woods were well marked with warning signs, and it remains a mystery as to why the couple were there. The dangers are widely known, as it is the third time this year that undetonated shells from the First World War have claimed lives in our district.

The next clipping was from the Saturday and reported that, at the time of going to press, I was still missing. A relative had arrived from Norway to identify the deceased.

Bestefar must have brought a photograph with him too, because the newspaper showed a picture of me in front of the storehouse at Hirifjell alongside another of a woman in police uniform. The caption said that her name was J. Berlet, and in the article she was quoted as saying that a team with specially trained dogs had been searching non-stop for the Norwegian boy since Friday, and that they had also dredged the pond in which my parents had drowned. The search had been complicated, she said, by muddy water and by the shells on the forest floor.

The last clipping was from the Tuesday, when everything was over. At least for the search crews.

> The missing Norwegian child was found Monday morning at a doctor's surgery in the coastal town of Le Crotoy. The police believe the child must have been abducted but will not go into the details of their investigation. They have made no further comment. Apart from minor injuries, the boy is unharmed.

I shuddered. It was as though everything had happened all over again. I had imagined the Ancre to be a large, clean river, like Laugen, but now the truth had emerged, and it was definitive. They had died in a muddy pond.

At first I tried to reassure myself with a harmless explanation; that Einar had heard about their death and ordered the French newspapers later. But the pages were crumpled and yellowed, with small tears along the edge, and the articles were heavily outlined with a pen.

He had been there. The question now was whether he had been searching, or fleeing.

Or had killed a family in France.

The woods north of Authuille. I had thought they had died in an open meadow, a battlefield preserved as an outdoor museum. But according to the newspaper it sounded like a dense and impenetrable place.

Maybe there had been more to the vague accusation Bestefar had mumbled at their grave. *Those damned woods*, he had said, and they were real. They must be close to Authuille.

I pushed these thoughts away. Eventually I found the key to the boathouse, cleared some space between manky coils of rope and unravelled fishing nets, and moved *Patna* inside. The weather changed constantly. The heat gave way to prolonged gusts of cold wind and rough seas, followed by sunshine and calm. Then a heavy downfall and more sun.

Occasionally a fishing boat passed, but always on the other side of the island.

I spent a long time simply standing in the workshop, following different trains of thought and exploring possibilities. One moment I was hungry for revenge, the next, I felt pity.

The priest said that Einar had become religious. *His doctrine was rock-hard and filled with remorse and anguish.*

But that was in 1967, I thought. Einar had been a tormented man long before he met Mamma, that is, *before* she died in 1971.

Greed of a kind.

I did not detect the slightest trace of greed here on Haaf Gruney. This place was like a monastery.

4

I RAISED THE OARS AND LOOKED BEHIND ME AT UNST. She must live near the coast, unless she was in the habit of going on long walks at midnight. *Here during the summer*, she had said. Lived at home, presumably. I had precious little interest in knocking on her front door and meeting her parents. Besides, I did not see any house which seemed likely, just the homes of old bachelors with diesel drums and crab pots in the garden.

Patna groaned. It was sickening to be using the boat that had crushed Einar to death. A couple of gulls followed me as I rowed across to the boathouse and pushed the vessel inside. I unlocked the Commodore, put on a pair of dry trainers and slung the Leica over my shoulder.

The weather changed again. Suddenly the daylight faded, as if wax paper had been placed over the sun. A milky fog settled in. It came and went, letting in blinding rays for a few minutes at a time, and I lost all connection with the landscape I was walking through. One moment I stood by a collapsed stone house, the next I stood on a narrow verge between a fence and the road. An orange Vauxhall rumbled past.

Now and then I raised the Leica and took a photograph. But I no longer had that desire, that need to capture the world on camera as if nothing should go to waste. Everything I encountered out here was so changeable, and it was almost as if I did not *want* to know what was real and what was not. I was reluctant to have the answers stored in photographic emulsion, waiting until I came home, and perhaps contradicting how I wanted to remember things.

Then I saw something, clear as crystal, glimpsed between breaths of hazy mist. A wooden house, the first I had seen on Shetland. Tall and wide, three storeys high and in a style that broke with its surroundings. Flat roof, large windows. Painted a light yellow, the entrance was framed by tall pillars, and at the end of a broad staircase, sheltered by a covered veranda, there was a generous set of double doors made of gleaming brown wood. A tall, sagging fence made of rusted wrought iron ran around the property. The view from the top windows must be formidable, as the house stood at the very edge of a cliff that plunged to the sea.

As I approached, I saw through the next break in the fog that the house was abandoned. The grass around it was tall and stiff. Two windows were broken, a side door was boarded up.

The iron gate creaked as I passed through it. Once a broad and geometrical grid of gravel had lain around the house. Now the grass borders were overgrown and unkempt. The stone steps had shifted slightly, and small plants were growing in the cracks.

The fog was dissipating now. I tilted my head back, looked up at the top floor. The house was so wide that it would have been impossible for the innermost rooms to get any daylight.

Something tingled inside me, a queasy feeling that my enthusiasm had got the better of me. I turned. On the hillside behind, a man and a young boy stood staring. Farmers. Wellies, raincoats, a sheepdog on a lead.

I nodded and held up the Leica in an attempt to bluff a reason for being there. But they did not react, just continued walking up the hillside.

How quickly rumours spread in a place like this. As quick as the wind and far and wide, out to everyone. All of them would pass a blue Norwegian-registered Opel Commodore. A dead man's rowing boat was set afloat again. Someone was nosing about on an island that had been inhabited by an *unken* body for decades.

And now that someone was snooping around here too. I went over to the wall of the house. The rumbling of the sea grew louder with each step I took, and soon I could no longer hear my footsteps on the gravel. By the time I rounded the corner of the house, the noise of the sea crashing against the rocks thirty metres below was deafening.

Then she was behind me.

"Where did you come from?" I asked.

She did not answer, simply signalled with her thumb that it was time to turn back.

"This is private property," she said when we were on the leeward side of the breakers. She was wearing a different jacket today, a rather tight, greyish-green tweed jacket with red lining under the collar, taken in at the lower back to accentuate her bum. She seemed in more of a hurry than yesterday, buttoning up her jacket as she walked.

"Whose house is this?" I said.

"This is no house. This is Quercus Hall."

"Qu—. What?"

"Quercus. Oak. The structure is made of oak."

"Do you live *here*?"

She shook her head. Kept walking, until we reached the gate.

"I'm just taking care of it," she said, shutting the gate behind us. "It belongs to the Winterfinch family."

I turned, did not want to lose sight of this immense, weather-beaten house.

"So where *do* you live?" I said. "Since you saw me coming."

She nodded towards a path in the grass. It led to a small stone cottage surrounded by a stone wall.

"Why didn't you tell me that yesterday," I said, "that you live with them?"

"It's not something one blurts out to strangers," she said. "I'm given use of the stone cottage and the boat in return for looking after the manor."

She wore an antique men's watch on her wrist, and she glanced at it impatiently.

"How did the rumours go about Einar?" I said. "More precisely, which year was it that he was supposed to have killed someone?"

"I don't know more than that," she said. "I'm heading to Lerwick. Have to catch the bus."

Just like the previous day she kept walking without checking if I was following, and just as before I shuffled after her.

Something was jarring with me, something she had said. A crackling fuse. It was like being in the mountains and being surprised by a huge reindeer out of hunting season, that feeling of excitement.

"I can drive you to Lerwick," I shouted after her, "if you like."

Her name was Gwen Leask and she wanted to go to Lerwick to buy the Runrig album "The Cutter and the Clan". She had grown up in the north of the island, but her parents had moved several years ago. As I understood it, the Winterfinch family would spend the summers here, and then she would prepare the house, clean and go grocery shopping. The only task she had at this time of year was to regularly check the entire roof with a torch and notify them of any leaks.

"So of course I can only do that when it rains," she said as we stood on the ferry. "And it does that every day. I like the rain."

"What do you study on the mainland?" I said.

"Finance. Numbers and figures."

Her sentences were truncated when she talked about herself. When I had trouble understanding her, she did nothing to tone down her dialect. But I was able to grasp that she lived in Aberdeen.

"What are they like?" I said. "The Winterfinch family. Why did they lease the island to Einar, and why have they left the house untouched?"

"Please understand," she said, "I can't talk about my employer. It's just not done."

We drove off *Geira*, and halfway to Yell the air grew dank and sticky. Without asking she twisted the knob for the fan to full blast, slid the seat back and pulled off her tweed jacket.

More of her body was visible now, pale and a little formless, but there was a sensuality about her when she twisted in the seat and I caught myself staring at her for too long. Her narrow eyes suggested she didn't give a damn what other people thought. She seemed sharp as a tack.

Where am I headed, I asked myself.

On board *Bigga* we bought chocolate from a vending machine. On the wall was a poster for a "Haltadans" on Fetlar, with Fullsceilidh Spelemannslag providing the music.

"We call it the same thing in Norway," I said. "Not that it's anything for me."

"Dancing?"

"Never," I said, and felt stupid again.

"You should have been here for 'Up Helly Aa'," she said and dug out a flat red pack of cigarettes. Craven As, with the face of a small black cat on the front.

She offered me one and I stood with chocolate in one hand and a cigarette in the other. Yes, she must be my age, twenty-five at the most. She kept her elbow close to her body when she smoked. With each drag her gaze looked beyond the horizon, and she moved her hand towards her shoulder so that the glowing tip pointed backwards, a posture which also necessitated a gentle twist of the hips.

She could even manage to smoke elegantly on a Shetland Islands ferry.

"What was that you said about 'Helly Aa'?" I said.

"New Year's celebration, the name comes from the Norse. Horned helmets and lots of beer. They build replica Viking longships, put them out to sea and set fire to them with flaming arrows."

"You said *they*. Do you no longer consider yourself a Shetlander?" I said, glancing around for an ashtray. She took a gentle step backwards to let the wind take the ashes. I tried to do the same, but the embers blew onto my anorak.

"I don't consider myself much of anything," she said, taking another drag before dropping the cigarette in the sea.

149

Bigga slammed into the pier of the main island. Gwen had thawed out a little during the crossing, but she kept trying to direct the conversation towards me and not herself.

"Is there any sort of registry office in Lerwick?" I said when we were back on the road. "You know, for title deeds and that sort of thing?"

"What do you want there?"

"To ask if there's any kind of agreement for Haaf Gruney. Since the buildings are empty, I'm wondering whether Einar actually owned the island."

"You'd have to go to the sheriff. There's no registry office on Shetland. No police either. The sheriff's office handles everything."

"Do you want to come with me?" I said. "It would be nice to have an interpreter."

"You speak perfectly intelligibly. Everyone in Lerwick enjoys meeting a genuine Norwegian."

"It is not actually the language I need an interpreter for. More to . . . to figure out how to ask."

"Oh please, I've said this already. I can't do that as I work for them. It is not done."

Every building in Lerwick had been lashed by rain; cloudburst we passed on Yell had passed through here first. Has there ever been a house on Shetland that was dry, really dry, I wondered as I headed for the sheriff's office on King Erik Street.

At the entrance I stood looking at the sign. Officialdom had always annoyed me back home, the staunch smugness. The Norway of skiers. The Norway of officer school candidates. But the Norway I found remnants of here was a warped and imprecise suggestion of everything I liked about my homeland. I had long since known that Shetlanders' Norwegian roots ran deep; you just had to look at the blue-and-white-crossed flag flying everywhere. But I had not expected there to be an Old Norse motto under the sheriff's coat of arms.

Með lögum skal land byggja.

I found myself at the counter of the archive. A map of Shetland covered the majority of the back wall. A man with a shiny bald head appeared, eating a currant bun.

"Yes, sir," he said. "How may I be of assistance?"

I went to the map and pointed to Haaf Gruney.

"I wanted to ask about the ownership of this island," I said. "A man named Einar Hirifjell—"

"I'm sorry?"

"H-i-r-i-f-j-e-l-l. He was a relative. Lived out on Haaf Gruney for nearly forty years. He may have got British citizenship after the war."

The man took another bite of the bun. Stared at a plastic flower in a yellow vase on the countertop.

"And now you're wondering who owns the island?"

"Yes."

He pulled out a threadbare hardbacked book, then walked over to a filing cabinet and flicked through the folders. He used both hands, with the half-eaten bun lodged between his teeth. The metal drawer slammed shut and he opened another. He tucked a folder under his arm and pulled out one more.

"Hm," he said, placing the folders on his side of the counter and chewing on the bun.

"Yes?"

"I have to ask for your name. Identification."

I handed him my Norwegian driver's licence. He studied it sceptically.

"You said your name was Edvard . . . *Hirifjell* – apologies if I'm not pronouncing it correctly."

"Yes. That's me."

"Do you have a middle name?"

"No."

"I will take that as a yes, because your identity number matches. You're to receive a copy of this deed. The right to Haaf Gruney was transferred from Mr Einar Hirifjell to Edvard Daireaux Hirifjell on November 5, 1971."

I felt blood coursing through every single vein in my body.

"In 1971?" I mumbled.

"Yes, indeed. But the transfer was not to come into effect until after his death."

"So I own an island?" I said, and looked at the wall map.

"Yes and no. Only on certain conditions. Here you have the original contract," he said, giving me a small piece of paper headed WINTERFINCH LTD.

It explained, in black and white on yellowed paper, that Einar Hirifjell and his descendants had permission to live on Haaf Gruney, and had exclusive right to the land and the buildings *until the end of time*. No rent was to be paid. The one factor that could overturn this perpetual contract was an *act of God*.

Beneath the line *As witness the hands of the parties* I saw Einar's steady, right-leaning signature. Next to it, much larger, the letters awkwardly scratched on the page with such a firm hand that the paper had torn: *Duncan Winterfinch*.

The agreement was dated August 3, 1943.

I looked at the man behind the counter questioningly. He glanced at the document. Ate the last piece of his currant bun.

"A rather generous agreement for its time," he said. "Nobody knew what the war would bring, and many feared that all of Europe would become German territory. A home in Great Britain, however meagre, was a ticket to freedom. But now it appears rather – how should I put it – characterised by the solitude of its location."

"What is an *act of God*?"

"Anything beyond human control. An earthquake. A volcanic eruption. The island sinking into the sea." He scratched his eyebrow. "But this is what complicates everything." He pulled out another document, with the same letterhead – WINTERFINCH LTD – but in a more modern style.

"A lawyer contested its legitimacy when Mr Hirifjell died. They said they would issue a writ if someone advanced a claim."

"Whose lawyer?"

"The Winterfinch family's, in Edinburgh. They believe the document to be invalid because Mr Hirifjell had no children."

"He's my great-uncle," I said. "I'm his closest heir."

"*Descendants*, according to this contract, means a child or grandchild, I'm sorry to say."

He went off to make copies and authenticated them with the sheriff's official stamp. A gentle thump on the stamp pad, a rough thump against the counter. A yellow plastic pocket to protect them from Shetland's perpetual rain. He offered a "Good luck" before waving me out.

I already had my hand on the door handle when he shouted to me from the other side of the room, holding up a piece of paper for me to see.

"This doesn't actually belong in our archives," he said. "Perhaps you would like it?"

I crossed the floor. "In 1971," he said, "when Mr Hirifjell wanted to transfer the rights to you, someone at the office must have helped him sort out the paperwork. Look at this."

There was a small sheet of graph paper with a pencil drawing of a round table. Below was a list of materials and measurements.

"Not that," he said. "On the back."

I turned over the sheet. A checklist. The writing was unsteady and haphazard, but it *was* Einar's. A list written by a ragged man on Haaf Gruney before an important visit to Lerwick.

Sheriff
Remember passport
Title deed
Letter to Edvard – that he must spend at least one cold week on
the island
Bank transfer to flower shop

I found her in the very back of Clive's Record Shop on the main street, flipping through the albums in the soul section. A record must have been set aside for her, because on the counter there was a sealed plastic bag with GWEN written on it in red ink.

I flipped absentmindedly through the albums, felt the weight of the vinyl building up in my hands before I tipped the pile back and moved to work on the next row. I couldn't arrange my thoughts. I found two maxi singles by The Pogues, but put them back.

Bank transfer. Had he meant the flower shop in Saksum? For Mamma and Pappa's grave? *Spend at least one cold week?* It made no sense. Presumably it was meant to be included in a letter to me, one that had been intercepted by Bestefar.

Gwen moved to a new row. Scratchy speakers were playing "Half a World Away" by R.E.M. She reached for an album, flipped over another instead and read the back cover. Quick and precise, she knew what she wanted. She ran a finger over her temple, sweeping a lock of her slightly curly hair back into place. She had an attractive back.

"Aren't you going to buy something?" she said, and pulled out an L.P. by Maria McKee.

I shrugged. "Don't have a—" I searched for the word as I spun my finger in a circle.

"Record player?"

I nodded.

"Hm," she said, as though opening and closing something simultaneously.

I bought a twelve-inch single of "Fairytale of New York", in the hope that it was also customary here for the nearest person with a record player to invite you home to copy the album onto a cassette. Later, as we were walking along the harbour, each with a shopping bag, she asked if I had had any luck at the sheriff's. I told her that Einar had transferred the island to me when I was three years old, but that the claim had been contested by the Winterfinch family.

"I expected as much," she said, offering nothing more.

I wanted us to drive out to Unst together, have her invite me in for a hot drink while her new record played on the turntable. I needed to be bolder.

"That's why I'm taking the first ferry to Edinburgh," I said. "I'm going to look up the Winterfinch family. Don't worry, I won't say that I met you."

She was about to say something, but stopped herself and walked on ahead of me. The car was parked behind Viking Bus Station, and as we approached it, an extractor fan sent an exotic aroma in our direction. An oily, delicate fragrance.

Raba Indian Restaurant.

We stopped instinctively. I caught our reflection in the restaurant window and thought, Dear God. It was only when I saw the two of us together that I realised how elegantly she was dressed. And that I would cut a hopeless figure at the door of a wealthy family in Edinburgh. I was wearing filthy black Levis, an anorak with bulging pockets and my hair looked like I had spent three days in the fields.

"Edward," she said. "Has it been as long since you last ate as since you last changed your clothes?"

I nodded. "I can't go in," I said. "Not looking like this."

She crouched by a drainpipe, filled her hands with water and ran her fingers through my fringe.

A girl looking after me. A girl who called me Edward. In a brief moment of afternoon sun. On a street in Lerwick.

"Take off your anorak and hold it under your arm. You look a little scruffy, but it'll be fine. This isn't Bibendum, after all."

"Bibendum?"

But she was already on her way in.

Five minutes later we each sat with a wine-coloured leather-look binder. Every time the door to the kitchen opened, and a waiter in a white shirt hurried between the tables, intense aromas drifted towards us.

She brushed something unseen from her cheek and got the waiter's attention. How did she do that? I had hardly dared to go inside, but she had two of the waiting staff attending to us as soon as we stepped through the door. Without a smile she had dismissed the suggestion of a table in the middle of the room, and instead pointed to a corner table that had just been vacated, with a request for "New linen, please."

When the table was ready and we had sat down, she said, "What would you like, Edward?"

"I don't know," I said, laughing quietly. I had to look at her to remind myself that we were the same age, that in fact she looked quite ordinary.

"What's so funny?"

"I've never been to a restaurant before. At least, not like this. Just the fish and chip shop in Brae. Great Britain's most northerly fish and chips."

"That's not a restaurant," she said. "More like a café."

The waiter arrived, a slim Indian with slicked-back hair. I stared at the menu in confusion. Gwen closed hers with a smack.

"Just a prawn korma for me, thank you. But the gentleman here would like mulligatawny soup for starters, rajasthani chicken for his first main course, lamb pasanda for his second main. And two peshwari naan, please, saag bhaji and tarka dal for the sides. Two pakora to share. Yes, poppadoms, of course, with some nice chutneys. Alcohol-free beer for him, he's driving. Give him a refill whenever his glass is empty, he's thirsty. A glass of red wine for me. Do you have a nice Barolo? And we'll need an ashtray, as you can see. O.K.? Lovely."

There must be a story behind a young girl wearing an old, scratched-up men's watch. Especially if she pulls down her sleeve whenever someone looks at it. I was certain it wasn't the only thing she wanted to hide.

"Why are you here?" she said. "Is it this fortune? The fortune that he, according to rumour, killed someone for?"

I shook my head. "No. I don't even know what it is."

"What *are* you looking for then?"

It stuck to me like a wart on skin. I had never really spoken to anyone about it, not even Hanne, whose eyes always looked elsewhere when the year 1971 came up.

Maybe it was because Gwen was a stranger, and the fact that I had the entire North Sea between me and the idle gossip of Saksum. Here was someone who was curious about my past, yet at the same time Gwen was like a rock lodged deep in the ground. I was tempted to coax the crowbar underneath, to see what would break loose.

"When I was little," I said, "I went on holiday with my parents. I got separated from them and was found four days later."

"I imagine they were happy to find you again."

"They both died. I grew up with my grandfather."

She had been rearranging the napkin on her lap, and now the stiff white fabric crumpled in her hands.

"Oh Jesus . . ." she said, then sat there for a long time before putting the napkin down. "You're alone!"

"Someone probably found me and didn't know what to do."

Thirty seconds ticked by on the unseen men's watch. I did not mention that Einar could have been connected to the incident, nor the newspaper clippings. When Gwen spoke again, she did not link my story to the rumours about Einar, as I had expected her to. Instead, it was as if she was processing everything in her head.

"There's no way it could have happened like that," she said. "That wouldn't have taken four days. Whoever found you would either have contacted the police at once, or had some reason for keeping you so long."

I looked at her. Here, for the first time, was someone who was interested in finding some *logic* in my disappearance, instead of burying it in some hazy oblivion and pretending it had not actually happened.

"Did you go missing first, meaning they died looking for you?" she said. "Or were you together, and you were the only one who got out alive?"

"The first seems more likely," I said. "Why else would they walk through a forest that was filled with unexploded shells?"

"Hm," she said, evidently still thinking.

"The strange thing is, we were there early in the morning," I said.

"Children wake up early," she said, and then added: "Or so I've heard."

I looked at her. "Maybe it was someone who wanted a child. I sometimes wonder what it would have been like, to grow up with another family and be none the wiser."

"You would have known," she said. "Sooner or later you would have felt it."

Like Mamma did with *her* adoptive mother, I thought.

"You don't remember anything?" Gwen said.

"Just someone arguing or screaming. I was sitting in a car. Something about a toy dog. It wagged its tail if I pressed a certain place. But I don't know if the memory is real or if I made it up. I was only three, well, almost four. I was born early in the year."

"I remember a lot from when I was four," she said.

The waiter arrived with the starters. Caraway flatbread dripping with oil, small ceramic bowls with orange and green sauces. I glanced at her, copying her method. When the flavours hit my tongue I was enraptured. Food with such complexity.

I suspected Gwen of hidden complexities too.

She had gone to the cloakroom to put on some make-up, barely visible, and changed seats when she returned. Now she sat with her back to the wall, on a banquette. The wallpaper was ornate and deep-red, and above her hung a painting of a tiger hunt with elephants.

"I still don't understand it," she said. "What are the chances of someone who would abduct a child being there, of all places? Even if the opportunity did present itself. It must be one in a million."

"There's something about that place," I said. "That place in France."

She looked at me. Quiet. Serious. Expectant.

The hell with it, I thought, *I'm going to hoist the flag*.

"During the war," I said, "my grandparents were notified that Einar had been shot in France, somewhere near to where my parents died."

"Where was that?"

"North of the Somme."

She broke off a piece of the bread and dipped it into the dark paste, which turned out to be aubergine chutney.

"Yes, but where exactly," she said, chewing slowly.

"Authuille. Do you know where that is?"

She swallowed exaggeratedly, signalling that the answer would soon come. "Of course. Everyone who listened in their history lessons knows about Authuille. It's near what's known as Blighty Valley, one of the most important British fronts during the Battle of the Somme. Authuille was bombed to hell. Have you been back there?"

"To Authuille?" I shook my head. "I've never been anywhere but here."

"Will you go?"

I fiddled with the ashtray. "I think so," I said. "Even though I had hoped to find the answers here. But all there is here is stone."

She looked away and I hastened to fill the void, so that it did not seem like a deliberate pause.

"And you," I said.

She just smiled, the same canny smile as when she made a comment about me getting married in Karmøy.

I had to stop myself staring at her. Gazing at her.

"And you, what was it like growing up with your grandfather?" she said.

I told her a little about the farm, and that the nearest record store was six miles away, six Norwegian miles. But just as she had her layer of make-up, I had my own mask. Saksum was infinitely far from the Raba Indian Restaurant in Lerwick. Everything I chose to reveal to her was filtered through a fine-mesh sieve, like

I was making juice with a tub of unwashed berries. By the time I had finished summarising the life of Edvard Hirifjell, the sieve was bulging, not with a mush of twigs, ants and pine needles, but with Front Fighters, silence and scowls outside the general store. I did allow the story about my mother to slip in halfway through, however. I had wanted to speak openly, for my own sake. Just to see how it felt.

I told it as if it was something I had known my entire life, that she was born in Ravensbrück and grew up in Reims, and that she came to Norway and was visited there by Einar.

"She changed her name," I said. "But I don't even know why she chose the name Nicole."

"What was the connection between them?" she said. "Between your mother and Einar."

"I don't really know, other than that he worked in France in the thirties. As a cabinetmaker."

"Do you speak French?" she said.

"A little. My mother spoke to me in French."

"*Il me semble que ce soit un bon souvenir,*" she said. That's a fine memory.

I cleared my throat, murmured the response to myself as I attempted to rediscover the melody of a half-forgotten language.

"*Oui, en effet. Mais c'est aussi tout ce dont je me souviens d'elle.*"

When the main course arrived, it was as if I had plunged into a hot bathtub and emerged with my eyes opened to a better world. Every fibre in my body seemed to be caressed and pampered. Small, dimpled pots on a brass stand with a tea light underneath. Swollen raisins swimming in a creamy sauce, sprinkled with coconut, skewers of reddish-orange, beautifully cooked meat. The waiter turned a bowl upside down over a white porcelain platter and revealed a dome of light-yellow rice mixed with grated carrot. Then two flatbreads arrived, straight from the oven, sizzling and glistening with fat and leaving behind the sweet, anaesthetising fresh-baked smell at the table.

160

I tucked in, and it was heavenly. Tastes and fragrances abounded, every mouthful was so good that I knew I would eat too much, and I allowed myself to. The morsels of chicken were spicy, I was sweating, and I wanted to sweat. It was like music building to a crescendo.

I looked at her across the steaming platters. She ate a little, and she ate elegantly.

Then the music changed. Until then it had been unmemorable, but now a song came on that I had despised back home. "Forever Young". Snobbish girls from Vinstra would play it at parties, rich kids would blare it from their new cars.

And I did not recognise myself; it was as if my shell had fallen off.

The song had a vulgar chorus with irritating sound effects and jarring echoes, but now that my defences were down it flowed into me, an intimate song, and I looked her in the eyes without saying anything, as if we were making a pact right there and then, a fragile pact with unknown consequences.

Let us die young or let us live forever.

It was plastic where real music was steel, a facade where there should have been a brick wall, but I heard it again. Passionately. I realised that this was one of those rare instances when music latches on to a moment. One I would remember for five years, if not ten. And I saw that she realised it too.

Then she disappeared.

This girl who said so very little about herself, who believed she had Norwegian roots – like most families in Shetland – and who wanted to return to Aberdeen "when summer is gone". As I scraped the last mouthfuls from the metal dishes, she got up without saying a word and gave me a hug. She mumbled that she was going to the bathroom, but the next moment she was standing outside in the street, with one hand raised and her fingers slowly waving goodbye. Then she slipped into a narrow alley and all that was left was the sheen on the paving stones.

I sat alone for fifteen minutes. Paid the bill. I didn't go looking for her.

Something had happened, something I could not decipher. Shetland was not a place where people raced out of their house as soon as a foreigner approached. But Gwen Leask had made sure to be in the vicinity, so that she could appear as if by chance when I arrived.

I walked along the empty streets of Lerwick and up to the small fort from which old cast-iron cannons pointed across the harbour. In the old days they had protected the herring trade, as an illuminated sign explained in both Norwegian and English. For once the weather was dry as I traipsed about looking through shop windows, at goods designed to withstand both the sea and the weather.

The city clung to everything old. Here it was still the chemist's job to sell cameras and film, a remnant from the time when people mixed the developing chemicals themselves.

I passed Hotel Kveldsro, at which Bestefar had stayed after the funeral. Wherever I turned, I saw traces of something Norwegian. Every second boat along the pier had a Norse name. *Nefia*, *Hymir*, *Glyrna*.

Back to the Commodore, my trusty, shiny-blue travelling companion. As I inserted the key in the lock, a thought occurred to me.

I soon found my way to St Sunniva Street, where a light shone in the flat above Agnes Brown's hair salon.

5

NOBODY ANSWERED WHEN I KNOCKED, BUT THE DOOR
was unlocked. I pushed it open and poked my head into the hall.
A grey raincoat, an umbrella. Women's wellies. The clothes of
someone who lived alone.

"Hello," I called, but there was no movement behind the cor-
rugated glass pane of the inner door. Further inside the house I
heard a faint melody. Someone was – humming?

I walked along the hall to a narrow staircase. "Anyone home?"
I heard the humming again, coming from above, where someone
was softly padding about.

I went upstairs and came into a narrow kitchen. The washing-
up had just been done, the room smelled of lemon and there was
a solitary plate steaming on the drying rack. A red kettle sat on
the stove, a cup of tea on an edition of *Møre-Nytt*.

"*Hallo*," I tried, this time in Norwegian.

The humming picked up again, apparently from another
room: "*Kjærlighet fra Gud*".*

"*Hallo*," I said, much louder now, and went back to the stair-
well, where I saw the slight figure of an old woman disappear
through a doorway.

Should I leave, come back tomorrow? No, she would be just as
hard of hearing then.

I followed her down another even narrower staircase, where
old magazines were stacked on the steps alongside the wall. I

---

* "Love from God ..." – a Norwegian psalm

emerged in a bare cellar and followed her through yet another door into a large room, where she switched on the lights.

The smell of burned dust stung my nostrils as I stared into Agnes Brown's disused hair salon. The sight that greeted me was nothing like the glimpse I had from in the street. But this was an interior I had seen before, in the catalogue from the *art-deco* exhibition in Paris in 1925.

The salon was lit by a row of rectangular lamps, like street-lights lining an avenue. The bulbs emitted a warm glow through orange glass decorated with frosted tulips bending on their stalks. Tall mirrors in front of each chair reflected the light. At first it appeared as though the floor was tiled in a checked laby-rinth pattern, but I soon realised that it was made of timber, with the contrasts created by slender blocks from different wood types.

Now the light was on her. Long white hair, a simple black dress. Surrounded by bottles of congealed hair lotion.

She went past the light-blue hair dryers into a corner where there were – six heads on a table? She ran her hand over each of them. Six plaster of Paris mannequin heads with wigs, the hairstyles straight out of black-and-white films. The humming stopped, as did the wistful, dream-like movements, and she picked up something from the table and tilted her head, just as Hanne had when she put in the earring. The hearing aid beeped when she had it in place. Then she spotted me in the doorway.

For an instant she froze, then took a step towards me and said in a Vestland dialect: *"Edvard. That is your name, isn't it?"*

Surrounded by the intricate designs, she wiped a little dust off one of the tulip lamps and gazed at me.

"I tried to ring," I said. "From Norway. But maybe you didn't hear?"

She glanced at the telephone. "Occasionally I come down to use it, to ring my sister in Måløy. Or to call the taxi service. But I do not want the telephone upstairs. I do not care for telephones."

The same was true for me, after all. "Was it you who decorated his grave?" I said, running my finger along the frosted pattern on one of the lamps.

She turned to the plaster mannequins. "Yes, that was me."

"I'm here to find out what happened in 1971," I said.

Agnes Brown had nothing to say to that. She appeared to be around seventy, and was beautiful in the way that a piece of expensive old furniture is beautiful.

"I am from Ørsta," she said suddenly. "Took my apprentice exam in Molde. I was Agnes Storeide before I married. He was a sailor from Lerwick, died in 1940. Torpedo."

"So it was you who told my grandfather Sverre that Einar was dead. Sverre is dead now too," I said. "I came to let Einar know."

"I thought that might be the case, that Sverre would have to die before you could come."

"Why did Einar have an entry under Lerwick 118 in the telephone directory?" I asked, stepping closer.

She seemed not to hear me. "What do you think it's like," she said after a while, "to wait twenty years for a call to come, and to top it all off it's not even meant for you?"

Agnes crossed the floor and adjusted her hearing aid. In a dark corner at the very back of the room there was a solitary stool made of enamelled white cast iron which shone in the street lights. "Gentleman's chair," she said. "Sit down. Then I can try to remember."

The dry leather creaked.

"Einar sat here, the first time I cut his hair in 1943."

Agnes Brown rummaged through a small chest of drawers. She turned on a verdigris brass tap whose pipes clanged as it spat out rusty water, and rinsed a pair of scissors. I looked at her curiously, and she grabbed a faded nylon cape and tied it around my neck. She put on a blood-red apron, placed both hands on my shoulders, and our eyes met in the crackle-glaze mirror.

"You look like him," she said. "I suppose that's not so strange."

Then, without ceremony, she began to cut my hair. Rapid, precise movements. Soon she put the scissors down, picked up a cut-throat and began to shear the back and sides, long strokes with the razor that made the roots quiver. She caught the hair in her hands, rubbed it between her fingers briefly before dropping it on the floor, as though judging its properties, and then began her story.

Einar had entered the salon in the spring of 1943, greeted her politely in decent English and asked for a *coupe Lyon*, with the hair set in stiff curls across the forehead. It had been popular in France in the late thirties, but Shetland was not a place where young men asked for fashionable haircuts. As a rule, they wanted their hair cut just well enough so they could remove their hat in church. At the time Agnes was one of two employees at the salon, which was run by an ageing barber from Glasgow. The furnishings consisted of peeling painted-wooden walls, rickety drawer units and a haphazard collection of mirrors.

Einar was so different from the fishermen who usually came to the salon. Lively and sinewy, familiar with French fashions and big-city habits. His accent was curious, and when Agnes realised that he was Norwegian, they switched languages without revealing their names. Shetland was teeming with Norwegians, and the young hairdresser and the light-footed man did as war conditions demanded, revealing little about themselves because there were ears listening all the way from Berlin.

She assumed that the hairstyle was a disguise for someone planning to travel to occupied France. Einar looked pleased with it as he strolled out with the door jingling behind him, and Agnes saw no more of him. When at long last the war was over, she took on the business, employed another hairdresser and concentrated on women's styling. Something happens when a hair salon expands from one to three ladies' chairs; it becomes the centre for local gossip, and soon Agnes overheard something that could only refer to the sinewy Norwegian.

One of her customers had been a cook for a wealthy merchant on Unst. She had been fired, and now she could indulge in slandering her employer. She said that he had invited a Norwegian to the house during the war, apparently a master cabinetmaker. The wholesaler and the cabinetmaker had found something in one another, not least because they both had a real appreciation of fine woodwork. Because the wholesaler was not just anyone. It was Duncan Winterfinch, fifth-generation timber merchant and head of a powerful family business in Edinburgh. They had moved the secretariat to their summer home on Unst for fear that their head office would be bombed.

Winterfinch and Einar had sat through the night in the drawing room, enjoying pre-war tobacco and forging plans. They had tea and sandwiches brought in at all times of the day. Eggs too, despite rationing. And nothing could create greater resentment than servants being forced to serve servants. They listened at the doors, tiptoed through the corridors, pieced together snatches of conversation which would be aired under a hair dryer on St Sunniva Street. It was clear that the two were devising some shady plan, because they went silent whenever anybody entered the room.

In 1943, with his newly styled hair, Einar had disappeared. In the months that followed, Winterfinch had become more and more uneasy. He would keep asking about a telegram that never arrived. He had always been a cantankerous man, not least due to old war injuries; one arm had been amputated, and bullet wounds in both legs meant that he had to use a wheelchair from time to time. It was during one such period in the autumn of 1944 – the pain was playing havoc with him – that he received a telephone call which infuriated him so much that he broke a window in his office. Nobody could discover what it was about. But immediately after the war, a similar scene played out.

The Norwegian had appeared at the door demanding to speak to Duncan Winterfinch. At first the butler did not recognise him; he was worn and frail, his body emaciated, his face drawn. The meeting had been brief. Winterfinch was furious, and within earshot of the servants he hurled abuse at the Norwegian, yelling that Einar had "broken the agreement and got the entire family killed".

He screamed so loudly that it echoed across the fields, and the next day all of Unst knew about the incident. Those words would cling to Einar for eternity.

What nobody had expected was to see Einar rowing out to Haaf Gruney, Winterfinch's property, and begin to use the houses there. Winterfinch had asked his butler to prepare a boat, he was carried on board and they set course for the island. The wheelchair was useless out there, so he had to be carried over to the stone buildings, growing ever more irate, and a fresh, bitter quarrel played out. Einar ignored Winterfinch as he would a hysterical child, turning instead to his butler and showing him a registered title deed stating that he had the right to live on Haaf Gruney, unconditionally and in perpetuity.

Winterfinch's retinue turned back with the matter unresolved; the one-armed man was so exhausted with rage that all he could do was gasp for air. But his reaction did not appear to have been provoked by the loss of a paltry sum of money. It seemed that Winterfinch had lost something precious, and he lapsed into a woeful, dejected silence, muttering something about "the poor widows". The next morning he fired the cook and three other servants, and for several days thereafter he kept to his office, from which all that could be heard was the squeaking of his wheelchair.

"This was right after the war?" I said, squirming in the chair.

"Yes, late 1945," Agnes Brown said.

So this couldn't have been about Mamma and Pappa, I thought, mulling over the discrepancy between what Agnes Brown had told me and what I had heard from Gwen. Agnes said that he had got a family killed. Gwen said that he had killed someone.

In the mirror I watched the hazy reflection of a car's headlamps on St Sunniva Street. A beam of light swept through as it turned, and I saw that my hair had been given a strange side parting.

"Did Einar ever talk about a woman named Isabelle," I asked.

"Isabelle, yes." Agnes laughed bitterly. "Isabelle Daireaux. Yes. I can promise you that he talked about Isabelle Daireaux."

One Saturday in 1945, around closing time, Agnes was sweeping the floor of the salon, and looking forward to dinner and a relaxing weekend with a novel sent to her by her sister in Måløy. Her attention was drawn to a gaunt face at the window, and she soon recognised the Norwegian whose hair she had cut two years earlier. His hair was tangled, his clothes filthy and he hardly seemed to notice the people around him. After everyone had left, he came inside and stared at the floor. Pointing at the telephone on the counter, he asked if Agnes could take a message for him.

"What kind of message," Agnes said.

"Isabelle," he said. "I have to get hold of Isabelle again."

"I'm not running a telephone exchange," Agnes said, putting down the broom, and then she told him about the rumours swirling in Lerwick that he had betrayed Duncan Winterfinch.

"No," Einar said, "Winterfinch betrayed *them*."

Agnes stared at the broom, and then at Einar, and thought that he both looked and smelled like someone she ought to sweep away. "My name is Einar Hirifjell," he said. "But if anyone rings, they might ask for Oscar Ribaut."

It was his desperation, intense and sincere, that Agnes gave in to. But she could not know that it was the beginning of a fatal, lifelong pact between the two of them. Einar explained, how during the war, he had travelled to the north of France via Spain under the alias of Oscar Ribaut. The aim was to carry out a secret, civilian mission for Duncan Winterfinch. Just what the mission entailed, he would not say. But the renumeration was to be a generous sum of money and the right to live on Haaf Gruney in perpetuity.

In France he had decided not to execute the plan, but his

reasons for that remained just as diffuse to Agnes as the mission itself. Instead he joined the resistance movement, La Résistance; he spoke perfect French after his years in Paris, and was accepted as though a born Frenchman. The resistance group, based in Authuille near the old First World War battlefields, was chronically short of explosives and so had made a reckless plan to procure more. They proposed to go into some fenced-off woods, gather undetonated shells from the previous war, remove the explosives – which are almost imperishable – and use them against the Germans.

This was how Einar met Isabelle Daireaux, my maternal grandmother. She was the oldest surviving child of a family of six. Two of her brothers had been in the same infantry platoon when the Germans broke through the Ardennes, and were shot within a day of each other. Einar moved into a shack on their farm, pretending to be an unemployed relative who had been a cabinetmaker in Paris.

As Agnes retold the story, it was unclear whether it was Einar or Isabelle who had suggested that they dig up the explosives, and thus bore the responsibility for the fatal events that later took place.

After the First World War, some time in the 1920s, the fields around Authuille had been ploughed up and cleared of skeletons and shells during a far-reaching campaign led by the French authorities. Several million tonnes of explosives were excavated, and the work was as dangerous as soldiering. Gradually the fields became tillable again.

But the battles in the woods had been so intense that there was no possibility of clearing them. There were more corpses and shells here than in any other place. Soon the groves were seen only as burial sites, best known by the nicknames the soldiers had given them. In this way, Bois d'Elville became known as Devil's Wood.

The occupying power knew of the remaining shells, but could not station a guard in the area at all times. Somehow Einar and

Isabelle had found a safe path into one of the woods, and began to gather explosives.

The explosives were to be used to blow a hole in the southern wall of a prison in the city of Amiens; the Germans had a number of central resistance figures imprisoned there. It was a decisive campaign, coordinated with the Allies and given the code name "Operation Jericho", for obvious reasons.

It was not long before Isabelle and Einar had fallen in love. Intensely, as Agnes had understood it, an impression gleaned from the few words Einar had used to describe her. The war made everyone quick to act, and greedy, because life could come to an end at any moment. Einar had acquired an audacity, and Isabelle too, and in that way they drove each other to be ever more daring. Soon Einar regretted not having transferred the farm to his brother. He considered it likely that he would die in France, and remain listed as missing in Norwegian records for years to come.

Only one other person knew that Oscar Ribaut was actually Einar Hirifjell, and that was Gaston Robinette, the leader of the resistance group. He was the head clerk at the agricultural bank – before the war an authority on counterfeiting, during the war an expert in creating false identity papers. Einar had kept his Norwegian passport hidden in the lining of his jacket, and had shown it to Robinette to gain his trust.

Einar found an unsentimental solution to the problem of the allodial right, which was to advance his own death. Robinette placed his Norwegian passport on the battered corpse of an informer, and in that way the message arrived in Norway which placed the farm in Sverre's hands.

The day of Operation Jericho arrived. The blasts were heard across Amiens, Allied planes dive-bombed the northern and eastern walls of the prison, enabling hundreds of prisoners to escape.

The Germans took their time with the reprisals and spent the following months rounding up La Résistance. People later believed that there must have been an informer at the centre of the resistance group, because one night, in the early summer of

1944, the Gestapo visited Authuille and the neighbouring towns. In a well-orchestrated raid with nearly four hundred German soldiers taking part, they apprehended most of the resistance members. Only a few, among them Gaston Robinette, managed to get away.

Einar was holed up in the woods that night, and he too got away. The Daireaux family was not so lucky. Isabelle, her sister, her parents and grandparents were locked up in that part of the prison that was still intact. The day after the raid Einar ran into Robinette and another resistance fighter in Authuille. They suspected him of having informed on the others, since he had so conveniently been elsewhere at the time. Robinette had pulled out a knife and asked whether the trick with the Norwegian passport had in fact been a signal to the enemy. Then, at the approach of a German patrol, Einar had made use of the commotion to flee. He left Authuille and called on a friend from his time in Paris, a man who now lived in a remote corner by the coast, and he remained in hiding there until the liberation of France.

Later Einar learned of the fate of the Daireaux family from a fellow prisoner. After being interrogated under duress for nineteen days, they were tethered to posts in the parade ground of the fortress, visible from the prison cells. The youngest girl, Pauline, was fifteen. For a long time nothing happened, the square remained empty, and then after an hour an audience arrived. Fellow prisoners forced by the Germans to watch. A soldier went to Isabelle's mother Nicole, and cut her loose, only to hang her with a noose suspended from a meat hook. When she stopped kicking they released the rope, took the noose from her neck as her body lay on the ground, and hung her husband, Edouard, with the same rope.

The rest of the family, the two grandparents and a cousin, were shot one by one. Pauline was spared, as was Isabelle, who was cut loose and tortured all over again. Some days later the sisters were sent to Ravensbrück concentration camp.

Einar was never free again. The worst of his torments was that Isabelle had gone to the camp believing that he had betrayed her. Directly after the capitulation he travelled to Germany and arrived at Ravensbrück, but could find no trace of her.

Winterfinch had smashed the window back then because Einar had telephoned to say that the mission had failed, and at the same time had asked for money to search for Isabelle.

To anyone but Einar it would have sounded like an illogical and illegitimate request, since he had failed to execute Winterfinch's mission. But when Agnes pressed him on this, Einar refused to say, until the day he died, what the original mission had been.

"So what do you actually *want*?" Agnes had asked Einar the day he visited the salon.

"To borrow your telephone number," he said. "So that she can get hold of me. Because I'm going to be travelling."

Einar would continue to search for Isabelle Daireaux, spreading the word that he could be contacted on Lerwick 118. At a well-run hair salon where someone was always available to take a message.

"But how will you fund your search?" Agnes protested.

"I'm going to help lift Our Saviour back on the cross," he said.

For Einar had a skill which came to be fully utilised and which provided him with shelter and the help he needed; a skill he had discovered in April 1940 when he pieced together the great crucifix in Saksum: he could repair church artefacts. Throughout the post-war period he travelled to war-ravaged cities around Europe visiting bombed-out churches, and offered to rebuild altarpieces and wooden sculptures that had been destroyed during the devastation. His only request was that the priests would look for any information about a prisoner named Isabelle Daireaux. He slept in the sacristies, repaired shattered furniture and ornaments, and in the evenings he wrote, with the help of the priests, letters to ex-prisoners of Ravensbrück, to the Red Cross, to the Allied offices, to the national registry offices.

Einar's one obsession was to find her, and he was far from alone; at that time tens of thousands were searching for their missing loved ones. Einar scraped together money for an old car, examined prisoner lists as soon as they were brought to light, and was met with a sad look of recognition when he visited the offices of the Red Cross. He told all of them that he could be reached on a Shetland number: Lerwick 118.

The same message was sent to Hirifjell in a short letter stating that he was alive, that he was relinquishing his allodial rights and had no wish to be contacted by the family unless someone died.

In the flat above the hair salon, Agnes soon made Einar's loss her own. She had his name entered into the directory with her number, as though they lived under the same roof. He called every week, from France, from Czechoslovakia, from the regions bordering the Soviet Union, and asked if anyone had come forward.

Before long Agnes began to hope that Einar would call just to speak to *her*. Whenever she picked up the receiver, she held out a flicker of hope that it was someone with the message that Isabelle was dead.

And so it went on, until Einar telephoned one day just before closing time. He did not hurry to ask the usual question, just said simply: "It's me."

Agnes had asked him to repeat himself, because his voice had completely changed. In a strangled voice he managed to convey that the Red Cross had found some documents from Ravensbrück; Isabelle had been sent on a death march, had frozen to death and was believed to be buried somewhere in eastern Germany.

For a brief moment, like a beacon of light through the hopelessness, Agnes was ashamed of the relief she felt. She hoped that Einar would now be able to forget Isabelle and return to Shetland. But that was not to be. Because Einar's despair was intensified by a piece of information given to him by the Red Cross, namely that Isabelle Daireaux had given birth to a daughter in Ravensbrück in January 1945.

As Agnes Brown cut my hair, strands kept slipping down my neck and my back and an itch had set in. At the same time an unease had grown inside me, an itch that could not be scratched.

At first I put it down to the certainty that the shells they had both collected were linked somehow to the incident in 1971. But then I felt an incredulity that ran much deeper, I felt it here and now as I sat in the barber's chair.

According to Agnes, Einar had not known if he was the father of Isabelle's child, or whether the child had been conceived as a result of rape. Since my grandfather had been fighting for the Germans, I could hardly deplore that fact alone; I could deplore the act of rape but not the uniform, because it was the uniform Bestefar had worn.

But my doubt soon evaporated.

Agnes had almost finished cutting my hair and was tidying up, mirroring her work in 1943 when she gave Einar Hirifjell a *coupe Lyon* and turned him into the man I recognised from the counterfeit passport.

At first I put it down to the hairstyle, the fact that the face in the mirror was no longer mine. But then I realised that the man staring back at me was Oscar Ribaut.

6

HIS SCYTHE WAS STILL SHARP. TWO STROKES OF THE whetstone and the rust the sea air had left on the blade was gone. I took it outside and cut the grass around the buildings on Haaf Gruney. Blue and black peat smoke billowed from the chimney. I had lit the kitchen stove and considered whether I should have soup or fried sausages for dinner.

If only we could have eaten together, Einar and I. We could have pottered about, become the realities of each other's assumptions. No need for grand declarations. The smoke of Early Morning Pipe in the air.

Even after Einar and Mamma had met, the uncertainty over who was her father must have ached inside them. But the truth had emerged last night – to Agnes and me – the answer to the most burning question of all. But it came twenty years too late.

They took hold of me, as though my entire life I had envisaged a living-room wall with the marks and hooks of pictures which had once hung there. Now their faces filled the frames: Isabelle Daireaux, Einar and Mamma; Alma, Sverre and Pappa.

My parents had been cousins. I felt no conflict, no shame, just a closeness. I knew that I would never tell anyone about it. Apart from the old priest, perhaps. Nobody else deserved to know, nobody else needed to know. It was not an issue; on the contrary, I was proud to have a bloodline connecting me to the altarpiece in Saksum, to Ruhlmann's Parisian workshop. But people would interpret it differently, I had seen that happen. Hateful people with long beaks hacking at the point where one was most vulnerable, the core of one's dignity.

Bestefar may have had an inkling that Einar was Nicole's father. He saw me grow up to resemble the brother he could not stand. Yet never a hurtful word. He just kept on with his seed potatoes, which were not related to each other, but *were* each other.

As I cut the grass I could feel Bestefar's presence, the man who taught me to wield the scythe. His warm hands over mine. The straight swing, my grip on the weather-bleached wood, the steel that left the grass lying in straight rows.

They were with me, all of them, I felt they were watching over me. Could I think of Isabelle Daireaux as my grandmother? There was a place inside me for that word. So far Einar was just Einar, even though there was another place for him too.

I had some food, kibbled Irish sausages, and tidied up, and by evening the house was warm and cosy. I put new batteries into the radio and listened to N.R.K. on a crackling long wave.

Persistent rainfall was expected across much of Østlandet. An unease lodged itself in my body. Too much moisture and the entire potato harvest could fail. As an additional reminder, a hard rain began to fall, so hard that the drops appeared to be coming out of the sea. The view of Unst was gone, and I feared that rain would be just as heavy back home.

The following morning I was up at the crack of dawn. The sea was flat and calm. I rowed across to Unst, sat on a hill and looked down on Gwen's stone cottage. At around nine I saw movement behind the curtains. She opened the front door, stretched in the fair weather and went back inside.

I checked the time and decided to go down, but was not sure how to explain my visit. Because I had learned something from Agnes Brown that suggested Gwen was not who she said she was.

*

When Agnes had finished cutting my hair we went upstairs to her living room. She gave me the set of keys to Haaf Gruney she had kept all these years, apparently relieved to be rid of them. She told me that Einar had searched for the child, whose name he did not know, for many years. He took no heed of how hopeless it was, nor did he know for sure that the child was his – even though this was now clear to us. Almost a thousand children had been born in Ravensbrück during the war, but the Red Cross estimated that perhaps ten or fifteen had survived.

I pictured him spending his days with C-clamps and wood glue, face to face with crucifixes, eye to eye with the apostles, hand in hand with the Virgin Mary. Driving out his demons. It must have been impossible not to become religious.

He had used the same method as in his search for Isabelle. Grateful priests helped him to write letters, he visited orphanages and enquired about every child born in January 1945. It was a search he must have realised was impossible, because even if the child were still alive, in all likelihood it had been adopted and was unaware of its origins. In the fifties he became more and more disheartened. Because of the Cold War, entry permits were difficult to come by, borders were closed, and Einar began to spend more time on Haaf Gruney, always restless, haunted, and with a hateful eye on Duncan Winterfinch. He fitted out the cabinetmaking workshop and earned enough money to survive by making simple furniture for a small shop in Lerwick.

"But where do you get materials from?" Agnes had asked him.

"It comes to the island all by itself," he said. Haaf Gruney was situated in a location where the ocean currents carried driftwood from both Russia and the Norwegian coast, and occasionally hardwood from America.

Agnes explained how Einar never lost faith that Isabelle's child would contact him, and they had an understanding that if anyone telephoned for him, she would go to Unst and paint a white X on the boathouse facing Haaf Gruney. Then he would take the boat across immediately.

A white X. I remembered the flaking white paint on the boat-house entrance. "Did many people call?" I said.

"No. In twenty years, not a single person telephoned for Einar Hirifjell. His sorrow was great enough to drag us both into the depths, and my only hope was that an answer would come soon. Even if it confirmed that the girl was also dead. But who was I to build my hope on the death of a child?"

Occasionally she glimpsed a sign of affection in Einar. She had often dreamed of fixing up the run-down hair salon, and one day he turned up out of the blue asking if she could close the shop for a few days. He brought in materials, covered the windows with brown paper and locked the door. Agnes heard him working long into the night. The next morning he arrived with chairs and a counter he had already made, and she saw the hint of a smile.

On the Sunday afternoon his hammer fell silent, and he led her down to a salon worthy of an exclusive shopping district in Paris. The interior was in a distinct and straightforward *art-deco* style, simply executed but filled with beautiful ornaments, and a row of handsome lamps to give the hairdressers good light to work by.

"This is me," he said. "The real me."

The most beautiful piece was a chest with thirty small drawers for scissors and other equipment. The face of each drawer was inlaid with white mother of pearl, and when they were all closed they formed the outline of a tulip, identical to that on the lamps.

"It is all driftwood," Einar said. "Like you and me."

It was the first time he had evidenced any feelings for her. Agnes embraced him, but it felt like putting her arms around a Vigeland statue. After dinner he had stayed the night, but the next day he had gone back to Haaf Gruney.

The furnishings in the salon became an attraction. She could not help losing herself to a man who created something so beau-tiful. For this reason she could never bring herself to sell the salon, and she got used to being a switchboard for a conversation that never happened. Each year Agnes renewed his entry in the telephone directory, each year she renewed the hurt of her own life being wasted along with his.

And so the years passed. Until 1967, when the telephone rang and Agnes Brown set down her scissors with a foreboding that this call would be different.

The line had crackled with an entirely different background noise to the usual local calls. A woman speaking terrible English in a formal, loud voice stumbled through every word. She appeared to be reading from a piece of paper, until she mentioned a name: "Einar Hirifjell." Agnes could hear from her dialect that she was from Gudbrandsdalen. "Speak Norwegian, for God's sake. I'm Norwegian as well," she said, interrupting the staccato reading.

"Tell Einar that Nicole Daireaux is here. In Norway. Back home at Hirifjell."

The person gave her telephone number and hung up, and Agnes was left holding the receiver without managing another word. Nicole Daireaux, Isabelle's mother, had she not been hanged in 1944?

For the first time in her life, Agnes failed to complete a haircut. She abandoned the customer in the chair and took a taxi to Unst.

Einar had hidden a tin of oil paint and a brush under a rock, and she painted the white cross on the doors of the boathouse. She stood for an hour in the gusting wind and rain and stared out at the barren island. Four times a day, to coincide with the bus schedule, Einar would take a break from his cabinetmaking, climb a small hill and look across to the boathouse, and on that day she saw him appear long before the bus was due. She had waved frantically, and before long Einar had arrived in the boat, having rowed hard. He stood there gaping when she told him where the call had come from.

"Hirifjell?" he said at length. "Nicole?"

He crossed the road and hurried down to the telephone box by the ferry dock, and when he emerged he was confused and distant. "I have to go to Norway," he mumbled.

Later she saw him in Lerwick, making for the Bergen ferry in the old grey car he owned. He simply disappeared. Not a word of thanks, not a telephone call explaining what had happened in Norway.

"I hated Einar at that moment," Agnes said, "hated him for not including me in something I had been involved with for twenty years."

She cancelled his listing in the directory and told herself he could go to hell. She saw nothing of him for four years, but heard from others that he came to Haaf Gruney now and then, and that he had once had a visitor there.

And then one evening in December 1971 she caught sight of him in Lerwick. He looked worse than ever. He had apparently begun to drink, and was staggering out of a seaman's café near the harbour, looking like a cormorant after an oil spill. Agnes knew that a man could go to ground in Lerwick in less than a month, and Einar appeared to be heading on a steady course in that direction. The gale whipped his hat into the sea as he roamed streets soaked by Atlantic breakers and torrential rain. In the end only Captain Flint's would let him in. While everyone else stayed at home in the days over Christmas, Einar lurched about on Lerwick's slick flagstones.

Once more Agnes gave him shelter. By then he was so filthy that it was difficult to see what was skin and what was clothing. Whether his afflictions came from sorrow or guilt it was hard to say. In the case of Einar Hirifjell, she could not separate one from the other. It was New Year's morning before he was capable of speech.

He told her then that Isabelle's daughter had travelled to Norway, believing that Einar lived at Hirifjell. Having cleared up what Einar called "a major misunderstanding", they had regularly exchanged letters and she had even visited him. He described those years as the best of his life.

But then Nicole and her husband had died in an accident, and this brought Einar to his wits' end. It was clear to Agnes that

Einar felt a huge sense of guilt over what had happened. But the only thing he wanted to talk about was the fact that her son – that I – was alive.

On January 2, 1972, Einar had rowed across to Haaf Gruney, and this time he was there to stay. He never touched another drop of alcohol, and for the rest of his life he built coffins. He had the materials brought over on the Bergen ferry, and through the seventies Shetlanders were buried in coffins made of Norwegian pine.

A few years later Agnes closed the salon and went to Norway to live with her sister, but her restlessness led her back to Shetland. She would row out to Einar now and again, observing each time how the white cross she had painted in 1967 became ever more weather-worn. She cut his hair, they had dinner, and she would stay overnight on Haaf Gruney before taking the bus home to the flat above her former salon.

When the boat tipped over and Einar died, he was put in a simple spruce coffin he had left for the purpose in the boathouse. He had already chosen a spot in the cemetery in Norwick, where the weather wore down the gravestones after only a few years. On the lid of the coffin he had carved OSCAR RIBAUT, as though desperate to convey the name to someone.

Agnes sent word to Sverre, who came to the Shetland Islands with a gravestone in the boot of his car, but otherwise showed no particular interest, neither in the island nor for Agnes' story.

Agnes Brown paused. I gazed at this elderly white-haired lady and wished I could do something for her. My thoughts strayed to Hanne. Perhaps her devotion was the same, greater than I was capable of recognising.

Then Agnes described the stranger who had turned up at Einar's funeral. An elegantly dressed girl of seventeen or eighteen, who came alone and seemed out of place. The girl pulled Sverre aside, but he shook his head and seemed as silent and as heavy as the gravestone he had brought with him. Agnes was wrapped up in her own grief and paid little attention to this stranger, but during the ceremony she detected something in the young girl's features. She had seen her many years earlier in Lerwick, hand in hand with an older man. At first she had been horrified at seeing the old man let her lead them out into traffic, but then she recognised him and saw that he had no choice. It was Duncan Winterfinch, the one-armed man, holding hands with his grandchild.

When Agnes rowed out to Haaf Gruney after the funeral to tidy up and clean, she discovered that someone had been poking around in the buildings. She changed the padlocks to a set she had bought in Ørsta, rowed her last trip on *Patna* and locked up the boathouse. She put the keys on Einar's old ring to have them sent to Bestefar at Hirifjell. As she left Unst, she walked past Quercus Hall and noticed that the girl at the funeral had moved in to the stone cottage.

As Agnes described her, I knew she had to be Gwendolyn Winterfinch, Duncan Winterfinch's grandchild.

7

"YOU'VE HAD A HAIRCUT" SHE SAID.

"Yes. At St Sunniva Hairdressers."

"Quite . . . individual. Nice, though."

She stood in the doorway of the stone cottage with both hands on the door frame. She was wearing a dark-green felt skirt with small checks and a high-necked black pullover. Her eyes flickered, as though she had focussed on something and had been interrupted. In a matter of seconds something had hardened inside her; she now stood silent and arrogant.

A cosy warmth leaked from the room through the open door, the kind of warmth a house can exude when a fire has been burning all night. But it swept past me and disappeared into the grey weather. She had everything on her side: the ownership rights, the language, maybe even the truth.

"Nice to see you again," I said.

She looked down. Then past me and up towards Quercus Hall, which towered on the edge of the cliff. She said nothing, as though embarrassed that I was there.

"Have they arrived?" I said. "The Winterfinch family?"

She said that they had not. Over her shoulder I could see into a small living room and heard a refrain from the record she had bought in Lerwick. I searched for the right words and said: "Can you let me inside Quercus Hall?"

"*What?* I would lose my job. They don't even like it when I have visitors here at the cottage."

I took a step backwards. So she was maintaining the bluff.

The cottage was faced with rounded stones that had turned light grey where the sun had warmed them. The surrounding wall was overgrown with moss, tall and thick, and offered protection against the sea wind. A tangle of flowers grew in the narrow garden between the stone wall and the house.

Still she did not invite me in, nor did she initiate a conversation. She radiated none of the heat of the room behind her.

"Duncan Winterfinch," I said. "When did he die?"

"Why do you want to know?"

"Because I've discovered that Einar kept something hidden from him." I looked her in the eyes and added: "The inheritance my mother was looking for. This must be what Winterfinch sent Einar to collect in 1943."

She shrugged. "I see. And who told you that?"

"Someone who knew Einar well."

I said nothing more. It must be a hazardous spot to be in, pretending to be an employee who shows no concern for a dead man's former obsessions. If she expressed any interest at all in the inheritance, her cover would be blown.

"Well," she said, "I suppose I could let you in *here*. Come on then."

She put a couple of thick, reddish-brown logs from a polished copper bucket onto the fire. The fire flared. Oak. It seemed that timber merchants did not burn peat, not even their "housekeeper", who did not seem to be taking care of the house. The cupboard doors were open, the bed was not made, there was washing-up to be done.

It was cosy and inviting, and I was on my guard. All that was missing was the smell of Indian food to make me sink into the same infatuated candidness as I had at the Raba. There was a deep, red-striped sofa in the living room surrounded by piles of music magazines. One of them had collapsed. *Record Collector*, *New Musical Express*. A double album was open on the table next to the steaming teapot.

"Where did you go?" I said. "After dinner."

"Had to catch the last bus."

She turned down the volume on the stereo. The record player was a Linn Sondek; a hi-fi enthusiast in Saksum would have to spend six months of benefits on one of those. A three-piece Audiolab amplifier. Two wide Quad electrostatic speakers. A system that befitted the five hundred or so L.P.s that filled the shelves.

"Are you a music critic?" I said.

She studied her nails. "Just interested. The records – they're not mine."

I took off my anorak. Before I rowed across, I had scrubbed myself with soap and changed my underwear. But even now I could invoke the tastes and aromas from dinner at the Raba. My skin still smelled different, from all the unfamiliar spices that slowly seeped out of my pores.

"Weren't you going to Edinburgh?" she said.

"Yes. But the ferry doesn't leave until tomorrow. Tell me, do you normally fly down there?"

"No, of course not. I take the ferry too."

"What's the Winterfinch family really like," I said. "They just seem like a big mass when you talk about them, like some kind of clan."

She began to straighten up the magazines, kneeling down with her bum resting on her shiny shoes. A skirt suited her better than trousers. The back of her hair was cut in a U-shape, leaving her black roots visible above the exposed nape of her neck. I felt an urge to stand close behind her, brush my nose against her hair to see if the subtlest of touches would overpower her brain, draw out her true self.

"They *are* a clan," she said, standing up with an armful of magazines. "An old family. But . . ."

"You don't want to speak about your employer," I said.

"Exactly," she began, "it's—"

"—not done," I said.

186

For how long would she keep up this facade? And why was she hiding behind it?

"So have you worked out what the inheritance is?" she said.

"I have, yes," I said.

We looked at each other, as if we were each focussing a Leica of our own. Either she hadn't realised I was lying, or she was a better liar.

"And you know where to find it?" she said.

I shook my head.

She put the magazines down on a shelf and stood with her back to me, her hands on her hips.

"But . . . you do want to find it?" she said after a pause and turned around.

I nodded. And I asked myself what price I would have to pay for that.

A little later I rowed back to Haaf Gruney and walked around the island until it got dark, on my own with the night and the sea.

Gwen and I had listened to "The Cutter and the Clan", and became once again who we had been at the restaurant: two false identities, but the only ones we could use with each other. She told me about her favourite bands, Runrig and Big Country, and when the needle lifted off the record, the silence weighed heavy in the room. We waved an awkward goodbye, said "See you around."

She must have seen right through my bluff, realised that I needed her to reveal more, but it was as though we wanted to start a game and could not find the dice. But I also noticed that it was easier this way. As long as we kept everything on a superficial level, I would not risk discovering the truth about my parents. This was about an inheritance – about money, gold bars, whatever it was. A small detour around the big questions.

A detour I had been taking all my life.

I strolled up to the bluff from which Einar had had a view of the boathouse. A sorrow from my teenage years resurfaced, when

I blamed myself for never having grieved properly for Mamma and Pappa.

I admired Einar's grieving, his reaction. He had suffered openly and genuinely in 1971, and he had taken on the torments that should also torment me. I was just like Sverre. He had said goodbye to my parents, got up from their graveside after breaking down, taken my hand, and then turned his back on them and got on with his life.

But the difference, I thought as I stood out on the rocky knoll, was that Einar had *known* them. My grief, no matter how hard I tried to rouse it, was no more a colourless blanket; it could apply to anyone at all, and so my longing was not true either. It became a snowflake in the air that drifted past without anyone noticing.

The next day I was back on Unst and drove to the telephone box. I realised it must be the same one Einar had used when he called home to Norway in 1967, as uncertain as I was now.

I dug some coins from my pocket, took a deep breath and dialled a number in Saksum. It was a number I had called many times, whenever things got tough.

Her father answered and I listened to his usual ambiguous "I see" when I said my name. We had one thing in common: we were happy to avoid speaking to each other. "It's him," he muttered, and I heard Hanne say "Oh" as she came on the line.

I was familiar with that "Oh" – it implied a great deal. At this exact moment it would mean *I'm angry with you and you need to suffer a little.*

"Are you back from Sørlandet?" I said.

"I didn't go."

"Einar died years ago," I said.

"Oh." Her tone softened a little. "Are you back home then?"

"I'm still in Shetland."

"Oh, come on, Edvard."

"I *am* in Shetland," I said and opened the door of the telephone box. "Can you hear the gulls screeching? I'm on an island called Unst."

She said nothing. There was a faint, electrical crackling on the line.

"This is the man who's hardly been to Oslo," she said.

I had meant to say that I missed her, because I did. The complication was that I did not miss all of her. I missed her warmth and her calm, but I did not miss the lead weight of her which threatened to hold me down.

"I'm managing just fine," I said.

"But why are you still there if he's dead?"

I wanted to tell her that there was a connection between Einar and the four days of my life I was missing. That I was trying to find out why Mamma had gone to Hirifjell, and what the inheritance was. But I sensed that that particular story was beginning to belong to somebody else, a girl I was both waiting for and was attracted to: Gwendolyn Winterfinch.

"Edvard," Hanne said, "I don't know where I stand with you anymore. I don't know what to think."

"Why do you need to think anything?"

"Shall I just hang up then," she said.

The telephone beeped. I put in another coin.

"Hanne, I'm sorry. That was a stupid thing to say. Can I ask you a favour? I'm going to have to stay another week. Can you swing by the farm and check for any signs of blight on the potato leaves? The key to the gate is under the black rock."

"You're asking me *what*?"

"Just to check for blight."

"And what if there is? You can't expect me to start spraying the crops and being your tenant farmer!"

"Just check to see if everything's O.K. And if you could return the library books on the kitchen table . . ."

"When do they need to be returned by? The library has a two-month lending period over the summer. How long do you plan to be gone for, one week, two weeks . . . three weeks?"

"Another week. The sheep are up in the mountains. And—"

"I don't understand," she said. "Why are you still there if he's dead?"

I drove to Lerwick, to the ferry terminal at Holmsgarth. It stank of fish and diesel. The man at the ticket booth had a double chin and appeared to have been sitting there his entire adult life without exercise. I managed to postpone my return ticket to Bergen, and at the same time asked the times of the ferry to Aberdeen.

It could be something to talk about with Gwen. I could keep up the pretence that I intended to visit the Winterfinch family in Edinburgh, press her a little more.

The man in the booth raised his voice and said that he would not be selling me a ticket to Aberdeen.

"What?" I said, leaning through the opening in the glass.

"No tickets for you, sorry!"

I was about to protest, and then realised that it was just a figure of speech. The reason, he said, was that they were expecting a storm. A powerful one. "There are strong gales coming in. Really strong."

Which is why they could not guarantee the arrival times. Or any arrival at all, it seemed. "Worst case scenario, the ferry can't dock," he said, "and has to wait out at sea. There have been occasions where they have had to wait like that for two whole days."

I drove away from the terminal, rolled down the window and held out my hand in the breeze. Breathed in the smell of grass, looked up at the clear blue sky. Where was this storm?

I was halfway across Yell when it started. A gust of wind jolted the car sideways, as if a spring had broken. The road turned down into a bay, and I saw the storm approaching out at sea. The sky darkened, as though night itself was on the way, and the sea was already high and white and frothy. Driving uphill there was such a headwind that I had to shift into third.

An hour later, taking *Geira* across to Unst, I sensed real danger.

The lorry drivers pulled up metal hooks from the deck of the ferry, grabbed thick orange straps and secured their vehicles so they would not slide around.

I was overcome by seasickness long before we got up to speed. *Geira* followed the swells like a float. Time stood still as the ferry was suspended at the top, before becoming weightless and falling straight into the troughs, with the water streaming over the bow and cascading onto the cars.

I counted three waves like that before I had to go below and throw up. When I came back up on deck, I saw that we were off course, Unst was not where it ought to have been.

We were heading out to sea.

In a cold sweat, I began to read the safety instructions. Lifeboats, muster points, evacuation procedures. The sea was greyish-green and beaten thin, like the wake behind a propellor.

In good weather, the journey took about fifteen minutes. After half an hour we were still not on course. But the crew were not shouting or showing signs of alarm, they just walked around with heavy steps, anticipated the drops when the boat fell through the waves, tugged at the straps on the lorries to check they were on the same notch as before.

The passengers took no notice either. People sat in their cars, switched on the windscreen wipers and continued to read the *Shetland Times* as the sea splashed over their bonnets.

I became a little less uneasy. Until an entirely new concern arose. Their actions could mean only one thing: that the weather could get worse.

At long last the metal ramp crashed down. I drove ashore, unsteady and dizzy, as though I had been drinking hard spirits all day. I parked up for ten minutes before continuing.

Through the rain I saw lights on inside the shop. A rare photographic theme, one that could exist only in this weather. But I had not been carrying the Leica with me for a few days. Perhaps a sign that I had begun to feel a little more at home.

I hurried inside and grabbed some sausages, tinned food and a bottle of White Horse. The shopkeeper seemed surprised to see me, and glanced furtively at the magazine shelf.

There she stood with her back to me, wearing a dark-green oilskin jacket.

I waited for her at the entrance. Her shopping bag was thin and light, mine was bulging with tins and threatened to split.

A boy from the Norwegian woods faces the storm with tinned food and distress flares. Gwendolyn Winterfinch faces the storm with six magazines, a box of chocolates and some looseleaf tea.

"Hello, stranger," she said.

"Did you walk here in this weather?"

"A little windy, eh?"

We sat in my car.

"The ferry to Aberdeen isn't running," I said. "I'll have to wait a little to visit the Winterfinch family."

A blast of wind made the windscreen wipers clatter. We drove towards Quercus Hall, and as we passed Einar's boathouse I looked at the waves.

"You can't row out to Haaf Gruney when the sea's like this," she said.

"I know. I was thinking of sleeping in the boathouse, or in the car. That's why I bought the food."

It hung between us, the possibility of waiting at hers until the storm had let up. The possibility of her revealing who she was. And I would reveal what I really knew in exchange.

"You don't have to sleep in the boathouse," she said when we were parked in front of Quercus Hall. "I can take you across. And pick you up again tomorrow."

"Is that a good idea?" I said staring out at the grey and frothing sea.

"I have access to an entirely seaworthy boat. But we'd have to leave now, before it gets really bad."

"Isn't it really bad already?" I shouted into the wind. "What if we wait a little and see if it gets better."

"It won't. It will only get worse. Look over there," she said. "The storm petrels are gathering."

A cluster of black and white birds could be seen through the deluge. They flocked together and settled on the wind, and they all headed towards Haaf Gruney.

"They can forecast the *real* storm," she said. "That's why they're called storm petrels."

8

WHEN I AGREED, IT WAS NOT BECAUSE I NEEDED TO GET back to the island. I did it to find out what could drive Gwen to defy the storm. To forgo the comfort and warmth of the stone cottage and her six brand-new magazines.

"Are we really going to use that?" I said, pointing at the jetty where the boat she had arrived in on the first day was being tossed about in the waves.

"Are you mad?" she said and led me towards a huge boathouse a couple of hundred metres away. "We're going to use this one," she said, opening the gates.

In the darkness I caught a glimpse of a slender boat. Gwen disappeared inside and the next thing I heard was the roar of a powerful motor. It sounded like one of the big American cars that cruised around the Mobil station in the neighbouring village back home.

She backed out swiftly. An antique speedboat, perhaps twenty feet long, with two rows of dark-red seats and a low windscreen. The brown mahogany hull was scratched, the varnish was cracked and dull, the gunwale and the bow clad with chrome strips that were speckled with rust. *Zetland* was painted on the prow in matt black lettering.

"What kind of boat *is* that?" I said, reaching for the hand she offered me.

"Have you heard of a Riva?" she said as I sat down beside her. The sea crashed around us. The motor sputtered as she warmed it up. "Well, this is a Riva, but from 1924, before they became too flashy."

The age of the boat and the elegant handiwork were visible in every nook and cranny. It was like taking a Rembrandt out in the pelting rain.

"You're sure it can manage sea this rough?"

In response she asked me to hold on tight, pulled back a shiny lever and raced towards Haaf Gruney. *Zetland* cut through the waves like a torpedo.

"Duncan Winterfinch supplied them with mahogany," she shouted over the roar of the motor, and when she revved the engine the exhaust sounded like an operatic aria. "He had this one built by old Mr Serafina Riva. The father of Carlo Riva, who in the fifties transformed the Rivas into the glossy Rolex boats they are today. When Rivas became flashy, Winterfinch removed the emblems. Or so the story goes."

She was proud of him, her grandfather. He made her open up. It was just a matter of applying a little pressure, then she would reveal who she was.

"Where does the name Winterfinch come from," I said.

"One of his ancestors was looking for a place to settle. It was winter, and he lit upon a lone finch in a tree. A migratory bird that had failed to travel south. It became the family's sacred tree. So the story goes."

Why didn't I just do it? Ask her to tell me her name, then I could tell her something in return?

In truth I was beginning to enjoy this game, and I could tell that she was enjoying it too. She liked to play "the other", and I liked it that the first card she played was the King of Spades. A card I could only beat if I showed her that I was holding the ace.

Her face was different out here in the wind. Her cheeks were flushed, her hair tangled. What held up were her clothes. The downpour had soaked right through my anorak, while she was dry under her oilskin. The rain was ice-cold and came down diagonally, the drops felt sharp. She turned to face them, let them prickle her skin.

"Saves money on facials," she said, unabashed.

The Riva was at full throttle and in a matter of minutes we had pulled up alongside Einar's rotting jetty. I tossed my shopping bag onto the boards and climbed out.

But instead of reversing, Gwen left the motor idling and looked across at the open sea, her hair flapping. She tossed a rope into the air so that the wind took it straight into my arms and asked me to drag the boat into the boathouse.

"Look at that," she said. The clouds were even darker than those I had seen on Yell. They were rolling towards us black and heavy, like smoke from a volcano. "Now it's getting really bad." She opened the boathouse gate and tied *Zetland* to some barnacle-covered posts inside.

"What would the Winterfinch family say to you using the Serafina Riva, or whatever it's called?" I said.

"Oh, Edward, they have another Riva, you know. Down in the Med somewhere."

"One of the flashy ones?"

"One of the flashy ones. They belong to a different generation. A different style."

"Where did you learn to drive a boat?" I said.

"My father." She snapped the metal latch into place. "That way, he said, at least it wouldn't be the weather that drove him from Unst."

"Then why did you move?"

"Because the weather drove my *mother* from Unst." She headed up towards the stone buildings while I stood looking out to sea. "Is it a storm coming, or a hurricane?"

"I have no idea. Those terms are useless out here."

"Is there anything worse than a gale?"

"A furious gale. We don't have any names for wind stronger than that."

We sat in the spartan living room, each on our own side of the paraffin lamp, and ate Jenkins' Cod Cakes as the wind howled beyond the walls. Even though I was sitting absolutely still, I could feel the movements of the boat in my body.

I waited for the right moment, and it came when she reached for the salt.

"I found some notes," I lied, "from 1958. It seems that Einar had suggested to Duncan Winterfinch that he buy back the inheritance for three thousand pounds."

Her hand was poised above the salt shaker, there was a twitch and then a hesitation. A fraction of a second during which her movement lost connection with her thoughts.

"I see," she said flatly. "And did they strike a deal?"

"It seems Winterfinch didn't want to," I said.

Then the mask was back in place. She went on eating and began to talk about the storm and the weather.

But it was plain that she knew some of the story. She knew what had and had not happened in 1958, many years before she was born.

I very nearly told her everything, but I had the dead to answer to. If she knew so much in advance, so that all she needed was a year or a place name, then she could go back and find what Mamma and Einar had been searching for.

The booming from the sea grew louder. It was raining heavily now, hard against the windows, and we edged closer to the fire. We said nothing more about Winterfinch or Einar, we kept up our acts as the bewildered Norwegian and the ignorant housekeeper.

This will not end well, I said to myself, the two of us here. Hanne so far away, distant in every respect.

I looked Gwen in the eyes and thought: How long can a girl and a guy be under the same roof in a storm before they sleep with each other? Every gust of wind pressed us closer together, the stone walls protected us from the storm, but something else was taking hold; she gazed at me for longer, and I at her, with a kind of conviction that we were cave dwellers who in the end would have to turn to each other to stay warm, bring children into the world, preserve the human race.

Her nipples were stiff through the woollen fabric. I was turning into an animal. Just as her eyelids were growing heavy and seemed to contain an invitation, there was the sound of shattering glass. The racket from the storm grew louder and clearer.

The enchantment was broken, we jumped to our feet.

In the kitchen shards of glass were spread all over the floor. The rain was splattering in through the broken window, the drops hissing on the cast-iron stove.

There was a knock against the outside wall, stone on stone. Then another.

"We have to fasten the shutters," she said. "The stones are blowing up from the shore and hitting the house."

"That's impossible."

"Seriously! The postmen here aren't allowed to stop near the shore during storms. Their vans get dented."

I went to the broken pane.

"Don't stand there!" she shouted, tugging at my sweater. "The shards will blow inside and cut you to pieces."

"So what do we do?"

"Do you think the shutters are for decoration? Go outside and swing them shut, and I'll fasten them from the inside."

I had to force the front door against the storm. When it was halfway, the wind caught hold of it and almost tore it off its hinges. The storm came in and blew the jackets along the clothes rail. There was a howling in my ears and I was soaked through even before I reached the first corner of the building. Around me it was as if the sea had risen several metres; it seethed and frothed grey-green, ready to swallow the whole island. I pressed on, leaning forward as though climbing a hill, turning my back on the wind to take a breath. Down by the shore there was a harsh rattling sound, like the reverberations when a load of rocks empties from a trailer, each searching for a place to settle.

I unhooked the shutters while the pebbles flew about me, and her arm came through the broken window and pulled the shutters to. We went around the house like that until it was secured on the outside and dark on the inside.

She came out and stood beside me. She was drenched in seconds, her hair clung to her forehead. Below us the waves twisted like immense wood shavings.

"You know Muckle Flugga?" she shouted. "The lighthouse on Unst? After only a few years in service, they had to build the tower higher because rocks kept smashing the glass."

"Soon the fish will be blowing up from the sea," I said.

She nodded slowly. "It does happen, yes."

The sea and the sky were dark and colourless. The two of us felt infinitesimally small before this fused, raging mass, it was impossible to see where it began or ended. The noise was as loud as a helicopter about to land.

And then I lost my balance. For a couple of seconds I truly believed that the entire island had worked itself loose from the earth's crust. But it was just that I had leaned into the storm, only for it suddenly to subside.

A moment later the waves calmed, as though realising that they were alone in a primitive battle and now the wind was gone. The sea subsided, whipped up and full of air bubbles. The air was still laden with rain, but glimpses of light appeared to swirl beyond the cloud ceiling.

"Is it clearing up?" I said, amazed.

The rainwater trickled down her flushed face, but she did not wipe it off. "No. That was a 'furious gale'. Now the one we don't have a name for is on its way."

The water gathered into a fresh stream that ran past our shoes. The roof tiles were still in place under the wire netting, but the door to one of the outbuildings had been left open, and was now crashing dangerously hard on its hinges.

"I'd better get some more peat then," I said. "While it's still possible to walk upright. And I'll fasten that door."

"Let's get going," she said, and with quick strides she walked down to the boathouse to check on *Zetland*.

I suspected that the cards I was holding were terrible, not least because the weather was on her side. I began to realise that the conduct of families long ago was not controlled by kind thoughts and considerate words, but by stone against wind. By cold steel against foes. By dry floors against flood. By the attraction of the frightened to the fearless.

In the outbuilding I filled the metal bucket with peat, hurried back and emptied it into the woodbox by the kitchen stove. The storm was building again. The sea grew higher and the howling increased. I ran back for more, to fill the tub by the fireplace, and passed the small shed near the main building. A thought struck me and I stopped, water running through my hair.

At home we had the woodshed near the houses, to avoid having to walk far in minus thirty degrees. So why had Einar stored his peat in the outbuilding that was furthest away? And stacked at the very back too? He would have had to squeeze between the tools and the generator every time he needed it.

"That should be enough now," Gwen said when I came back in. She had lit a row of candles and sat in front of the fire wearing a thin grey-and-white-striped jumper. Through the material I could see the fastener of her bra.

"A little more," I said. "This storm could last."

She gave me the same languorous look that I had almost succumbed to before. "That," she said, "could easily happen."

I replenished the fire and went back to the outbuilding. The peat blocks were greasy and had a pungent smell. I dug down in the pile, moving them across the floor, and came to an older layer. The blocks were cut more evenly and the surface was grey and dry. Eventually I reached the floor, then kneeled down and swept my hand carefully across some wooden boards.

The floor? There was no wooden flooring in this outbuilding.

I remembered I was to *spend at least one cold week on the island*. So I would get through enough peat to reach the bottom.

*

"What took you so long?" she yelled.

"The door came off its hinges," I hollered. "Had to screw a board over the door frame."

We shouted even though we were indoors, because now the wind was bellowing down the chimney, like a giant blowing across the top of a bottle.

"Did you *screw* the door shut?" she said.

"It would have blown away otherwise."

I pulled off my soaked sweater and stood there bare-chested. "We have enough for two or three days now."

Gwen walked towards me, her body in silhouette against the glow from the fire. My jeans clung to my thighs.

"What is going on with you, really?" she said. "Did you see a ghost out there?"

The house began to creak. The squalls dashed it with waves, as she had predicted. Water began to run from the windowsills.

"This house is more than a hundred years old," she said, laying a hand on my chest. She took my sweater and hung it over the back of a chair.

"So if it hasn't already blown into the sea, it won't happen now?"

"Exactly. Great Britain's most powerful gust of wind was measured on Unst. At Muckle Flugga, in fact."

"Was it worse than this?"

"173 miles an hour."

"That's impossible," I said and wiped the moisture from my forehead. I had to get closer to hear her.

"The meteorologists said the same," Gwen said. "The measurements went off the scale. Then there came a squall that was even stronger, and it blew the measuring instrument out to sea."

The shutters rattled. We stood in the half-dark, in the warmth of the peat fire. She took a step towards me and bit her lip.

"But this house is still standing," she said.

She wanted me to take the next step, so that she could open

up and let me in. I could smell her, her hair and the wool of her undergarments, the smell of the sea and wet animal.

In the morning the island was damp and scrubbed clean. The sun broke through and the ground steamed, a strange combination of salt and earth, of rain and rot.

Further inland there were black strips of earth where the storm had torn up the turf. Floating in the sea were clumps of grass and seaweed. Driftwood had washed up and lay flecked with foam. Yellowish-white wood shone where the bark had been scraped off. It was mostly pine, Norwegian probably, but there were also some rough trunks that I did not recognise. They could have drifted from any of the seven seas.

There was Gwen. She took no notice of the fair weather, walked right past me down to the boathouse. Scorned and angry.

I had nearly given in last night, nearly given in to desire and curiosity, both for who she was and for what Winterfinch had been so keen to find – two questions that met at one and the same point. I had wanted to as well. To see her naked shoulders, tear off the rest of her clothes and take her on the floor.

But a cynical, almost wicked common sense had stopped me. There was something under the boards in the outbuilding, so well hidden that Einar must have had good reason to hide it. Something Gwen was probably searching for too.

Now she was marching down to the edge of the sea.

"What's wrong?" I said when I caught up with her.

She did not respond. I saw a dead sheep bobbing not far from the shore, its legs stiff. Now and then the waves took hold of it and turned it over, so that its hooves stuck out of the water like burned matches. Its head flopped loosely and its tongue dangled out of its mouth.

I waded in and grabbed it by the hind legs. Its fleece swayed in the water, first following the movements of the sea, and then mine as I dragged it ashore.

"What are you going to do with it?" she said.

202

"Do with it?"

It was a sheep. A farm animal. When I brought it ashore, I felt as if I were bringing ashore a piece of my life on Hirifjell. Heavy, exacting. The weight of my own ties to the farm. The sheep had a yellow clip in its ear, the same colour we used back home.

I pulled the carcass up onto a large, flat stone. The water ran slowly from its wool. Gwen strode off towards the boathouse, as ill-tempered as the after-effects of a "furious gale".

I expected to hear the sound of the metal bolt and the creak of hinges, but instead there came a loud, coarse "What the hell . . .?"

The storm had torn open the gate of the boathouse, battering *Zetland*. Part of the stern had been shattered, the wood was splintered like an old broken ski. But it had not taken on water and still floated as it should. There was a small rainbow-coloured slick of oil in the water.

She hopped on board. The motor started at the first attempt and she reversed out. I walked along the rocks to where I could climb on board and wondered if the boathouse on Unst – where *Patna* was moored – had survived.

But she did not bring the boat towards me. She revved the motor with the boat pitching a few metres out, studied me as Winter-finch must have studied Einar, and said "Goodbye, you—"

The rest of her words were eaten up by the roar of the engine. *Zetland* levelled out and she disappeared around the island towards Unst, leaving only her wake.

I left the sheep where it was. I unscrewed the boards that covered the door of the outbuilding and pulled away more of the peat. My hands were filthy. Soon the pile was so tall that it blocked the fresh air from the open door, but now I had dug down to a long, black object.

A coffin.

Outside, gulls were beginning to venture forth after the storm. They circled above the island, and through the open door I could see the bravest of them nipping at the sheep. I found a knife and went out to skin and gut the animal. The blood washed the black peat off my hands. I hung the carcass in the outbuilding and threw the entrails into the sea. Gulls swooped and squabbled.

I listened for *Zetland*. Nothing. The screeching of the gulls grew louder, then died down when there was nothing left to eat.

I went back inside to examine the coffin. In the light of a paraffin lamp I wiped the surface clean. Lilies appeared, gently looping lilies carved into polished black wood. The corollas and the stalks were made of glimmering, white mother of pearl. Light played on the ornamentation, like a low sun on a landscape, and further patterns appeared. The long sides depicted a forest. Not some flamboyant motif, just elegant, delicate etching of large solitary trees, peaceful and strong, among clusters of small ones. Blades of grass on the forest floor appeared to have been drawn freehand, impossible even for an artist with a sharp pencil. And Einar had used a gouge.

The coffin now stood free before me, surrounded by clumps of peat, like a casting mould that had crumbled away.

The truth. Once he buried the truth here. I grasped the lid, but it seemed to be stuck. I readjusted my grip and pulled so hard that the entire coffin lifted with a groan. Then I felt something give, the lid of the coffin creaked and came loose.

I kneeled with the lantern in my hand. Relieved, and at the same time disappointed.

Inside there were two objects.

An old, slender shotgun. At close range it looked almost hairy; the long barrels had been treated with grease and were covered with thick fuzzy dust. The woodwork was coated with a sticky wax that turned the stock grey.

The second object was a chest made of polished flame-birch. It appeared to have been cut from a single piece of wood; only when I took it into the workshop and turned on the electric light could I see the thin line of the opening. The chest was so tightly constructed that I had to use a screw clamp to get purchase, and when it creaked open, a puff of air nearly knocked me over.

It smelled secure. The smell of home. Of Mamma.

Inside was a soft package covered in grey tissue paper, and beneath it some letters. Addressed to Einar Hirifjell, postmarked between 1967 and 1971.

The handwriting was Mamma's. Thin airmail envelopes with ninety øre stamps, posted in Saksum.

As I picked up the package, the wrapping paper slipped off. The contents had been neatly folded and were now spread over my arm.

A dress of a deep navy blue, with white edging at the collar. A scintillating colour and a material of real quality, just as delicate after all these years in an air-tight chest.

It was like receiving an electrical signal from far away. Was I remembering correctly? I sensed a closeness and a warmth, and I recalled movements in a blue colour, but I could not be certain. I put my nose to the fabric, tried to recover the fragrance. Held it up in front of me.

Was this my mother's summer dress? The cut looked foreign, but I really could not tell.

Because it did not contain Mamma.

Everything around me had disappeared. Sounds which before would have been warnings did not trouble me now. The weather did not exist. I was sitting on a flat rock staring at nothing. The rain came, I got wet and dried again where I sat.

I wanted to get away. The sea was no longer merely a shield from the outside world, now it held me prisoner on Haaf Gruney too. Einar himself had become a ghost wandering aimlessly ahead of me.

The coffin. A slender, octagonal design. More beautiful and more bleak than anything I had ever seen.

He must have made it for my grandmother, for Isabelle Daire-aux. In the hope that her remains would one day be found. The trees were the front line, the lilies a bridal veil. The woods in Authuille, presumably.

I looked at the shotgun again. Beneath the dusty wax I saw that the stock was made of walnut, as was often the case with Einar's gifts.

The evening was approaching. I went down to the workshop and sat with the dress across my knees. A faint recollection stirred, but it would not emerge. It was like standing in front of a locked door, with the memory making a racket inside, while both the memory and I searched for the keys.

I closed my eyes and lifted the dress. Ran my fingertips over the fabric and the seams. Noticed that its texture was stored inside me.

A vision of Mamma emerged, with me hiding behind her legs. I came to just above her knee and pressed myself against her. I could smell suntanned skin, a bright sun shone and dappled me with blue. I saw some large trees and heard voices I didn't recognise, and I realised that one of the voices talking was me, it was summer and I said something to my mother in French.

I opened the first letter.

9

HE IS A RESTLESS CHILD, ENERGETIC AS A PUPPY, AWAKE AT first light. And then all he wants to do is hide behind the apple trees. And no matter how tired I am, I join in, because every time it is as if we rediscover each other, it is a reminder that life has meaning now. I speak to him in French, Walter in Norwegian. We wonder what his first word will be. Bet two kroner on it being French.

Mamma and Einar had exchanged letters in French. Her handwriting was uneven and twisting.

My dream was to become a glass-blower, she wrote. *I was not particularly good in school, but I was skilled with my hands, and I had good prospects of working as an apprentice.*

If only she had become that, I thought. A glass-blower. Left behind something permanent, evidence of a skill to create objects of beauty, something to remember her by.

I felt a cold shiver down my spine, followed by an uneasy kind of affection. She described me as *la lumière forte et belle* – the bright, beautiful light of her life. Who had saved her *from the darkness.*

The *darkness*? I put the letter back into the envelope, sorted them by date and began with the first. In it she mentioned, in somewhat abashed words, her first meeting with Einar in Norway. It seemed she had insulted him, called him something nasty, and she referred to this as a "misunderstanding" which she was now apologising for.

You have to understand that I came to the farm with thoughts of revenge. Strange to think about it now. Instead I found a home. The

207

traitor was not there. Just his brother, who said that you were dead. I forgive him for lying, because it was his way of giving me strength. I soon understood that he too had been hard hit by the war. It was strange to be telling my story to a man who had worn the German uniform. Recount every detail my adoptive mother had been through in Ravensbrück. Tell him that my true mother had died because of his brother.

Sverre told me that you were a dreamer who was not aware when you upset the lives of others, or put someone in danger. Alma was and is ill-tempered, uncommunicative, for reasons I do not understand. She is content with the practical things in life. Sverre himself said that I brought a new light when for him all light had faded. Then you came and threw a spanner in the works.

It ended there. She did not even sign her name. Einar must have sent her a reply straight away, because only twelve days later there came another letter from Shetland, stamped at the post office in Saksum.

From then on her tone was calmer, more familiar, and she began to talk about her childhood. I wondered what Einar had written that had put her mind at ease, and it struck me that his letters might still be somewhere at Hirifjell.

I would like to visit you, she wrote. *Has to be soon, if so. I am in my fifth month. Later it will be difficult to travel.*

In January 1968, a postcard was sent from Lillehammer hospital: *Beautiful baby boy! The name has to be Edvard, after my grandfather, but the "E" you can consider yours.*

Mamma's words both stung and consoled me. I read the letters several times, learned how she expressed herself, perceived what was hidden in what she did *not* say. She had written everything that Francine Maurel had told her about the women's camp. In this way I could piece together the stories of my mother's and grandmother's lives and, finally, the reason we went to France in the autumn of 1971.

*

In 1944, having seen her family hanged, my grandmother, Isabelle Daireaux, was crammed into a goods train and sent to Ravensbrück. Her fifteen-year-old sister, Pauline, died on that train. Isabelle had held her thin corpse until rigor mortis set in. The guards forced her to leave her sister in a pile with the rest of the dead.

In the camp she got to know Francine Maurel, who had arrived a few months earlier, and gradually found her own ways of surviving in this moonscape of cruelty and suffering. My grandmother must have been strong, but she broke down when she witnessed humanity reaching rock bottom in this place. Most of the children born in the camp died within hours of their birth; some were killed by the guards, others died in the gas chambers or as a result of enforced abortions. If after all this the guards heard a child's cry, they had a new candidate for their medical experiments.

Since my grandmother had been a member of the resistance, she was a *Nacht und Nebel* prisoner. One who would disappear and die. She was denied aid parcels, and was assigned to heavy labour in the laundry. Their meals consisted of a foul, brown gloop that was supposed to represent soup, and she soon became very thin.

It was a shock for her to discover that she was pregnant. Not the thought of having a child, but of the kind of world this child would die in.

The older children, the ones who had accompanied their mothers at the time of their arrest were now five or six years old. The children understood what was going on, and adapted their games; instead of cops and robbers they played S.S. guards and prisoners. They ordered one another to do forced labour, and did not hesitate to inflict a blow if the tasks were not properly carried out. Later they began to set up cardboard boxes, to pretend they were sending their playmates to the gas chamber.

All the while the rumours spread. More women from the resistance movement in Authuille arrived, and passed on the story about a certain Oscar Ribaut who had informed on them.

Perhaps my grandmother did not want to admit that she had been with him. In any case nobody knew who the father was early in 1945 when she gave birth to a baby girl in a corner of the laundry. Francine Maurel stuffed a towel in her mouth to stop her screams, and the two managed to wrap the baby in dirty clothes and hide her from the S.S. guards.

Francine had also been pregnant, by a German guard who gave her food. But this child was trampled to death shortly after his birth. Francine was better nourished than my grandmother and was able to give this new baby her milk.

A few weeks later, as the Russian army drew near, thousands of women were driven out of Ravensbrück with no more than the clothes on their back. It was a wretched and rainswept column. Barely a sound could be heard; they were too exhausted to whimper, and they did not know where they were going. My grandmother was wearing the same pair of shoes as she had worn at the beginning of the war. She had pneumonia and was coughing blood. Francine guessed that she weighed no more than forty kilos.

They took it in turns to carry Mamma. Two or three days after they left Ravensbrück, a group of women made a camp near a river and a German family gave them food. When Francine woke late in the morning, she saw Isabelle lying dead on the ground. She had wrapped Mamma in her prison uniform and she froze to death. Francine picked up the baby and felt that she was still breathing.

The farmer who had given them food dug a grave beneath a tree. The women began to sing a hymn, but most grew impatient after the first verse and trudged on. Francine took Isabelle's prisoner identity card, stood with Mamma in her arms and sang to the last verse. The farmer wedged a stick under my grandmother's back and rolled her into the grave. Francine caught up with the rest of the column.

She could vaguely remember them arriving at a village with a burned-out church, and meeting the white buses of the Red

Cross which were to take them to Malmö to be registered. Francine said that the child was hers and Mamma was christened Thérèse Maurel. She grew up in a tiny flat in Reims, and in time she wondered why she did not look like any of her relatives. In a Christmas clear-up she found a stranger's prisoner I.D. card, and Francine broke down and told her everything.

In January 1965 my mother travelled to Authuille and asked for directions to the Daireaux family home. The only thing she found was their memorial stone, and those who had appropriated the farm took her for a fraudster and told her to leave.

Mamma had nothing with which to prove her claim, and anyway, she was ten years too late. In France, 1955 had been the last year in which it was possible to demand the reparation of property forcibly abandoned during the war. She had no pictures to prove a family resemblance, and as she walked away her desire for revenge was irrevocably stirred. She commemorated the names on the gravestone and decided to change her name to Nicole Daireaux.

On her birth certificate was written FATHER UNKNOWN, and Francine repeated the theory that a German soldier had raped Isabelle. But another rumour had persisted over the postwar years: that there had been an informer, the one called Oscar Ribaut. Mamma talked to survivors of the resistance movement in Authuille, and met Gaston Robinette, the man who had seen Einar's passport. This supported the story of his deception, and she discovered that Oscar Ribaut's real name was Einar Hirifjell.

Mamma had not gone to Norway for a holiday.

She had gone there to settle a score. To show Einar her face, confront him with the fact that her family had been executed. On borrowed money she travelled to Norway and found her way to Saksum.

But she did not find Einar at the farm – only Sverre. And if there was one thing Sverre Hirifjell was good at – as good as his brother was at repairing shattered altarpieces – it was repairing the wounds of war. He kept up the story that Einar was dead, and said that a good life was yet a possibility for those who turned to working the black earth and to the comfort of Handel's organ concertos.

Perhaps he became a father figure to her, a benign version of the unknown German soldier she had come to believe was her parent. Even Sverre was certain that Einar would never return to Hirifjell, and he maintained the more palatable untruth about him being dead, an untruth that protected both Einar and Mamma.

The strange French girl arrived in the middle of the lambing season. A busy time. She stayed and helped out, despite Alma's scowls, and then Pappa came to visit. Someone her age, holding out his hand to others who had been born during the war.

Alma had seen Einar in the early years of the war, and perhaps now recognised something in her face. With growing unease she found the telephone number in Einar's old letters. She sacrificed peace on the farm and rang Lerwick 118, to prevent two cousins from getting married.

But by then Mamma was already pregnant. Again, blood had taken precedence.

In the letters there was nothing about Einar being her father. But I sensed a great intimacy between them after her visit to Haaf Gruney. Perhaps a touch was enough, a knowing look.

In the last letters one word kept recurring: *L'héritage*. The inheritance.

It seemed that Einar was the first to broach the subject, and he tried to convince her that she ought to resolve the matter. From Mamma's response it appeared that the items of value were in France. But she seemed to resist her right to claim them. For the first two years after I was born she seemed hardly to leave Hirifjell, except to visit Francine, who was ill and had not long to live.

Edvard is my life now, and I have no desire to see France again for a long time. Though perhaps it would be the right thing to do, to see it through. That is, if the other party is amenable.

Then came the turning point. It meant that I could no longer read the letters as a conversation between two dead people, but as the start of a story about myself, one that had led me here to Haaf Gruney, and one that would not end until I returned to France.

In July 1971 Mamma wrote:

Let's do as you suggest. I sense that you will not find peace, and I myself feel it more and more, this unease. Yes, I suppose we will return to Authuille. Walter believes that September would be a good time, it fits in with the farm work. I told Sverre that we wanted to go on holiday. That was how I put it, you know how he has such a hard time dealing with the past. He told me I could borrow the car! The lovely black Mercedes.

We will give the one-armed man just one chance. If he does not accept, I will shake hands only with the hand he is missing.

Yours, Nicole.

10

"WHAT HAPPENED? AND HOW DID YOU GET OVER HERE?" she said.

The same scene as last time: the warmth drifting through the open door, me standing cold and uncertain on the front step. I had hitched a ride with a fisherman and was more windblown than ever, wearing a wrinkled shirt and trainers with wet grass stuck to them.

"What's that?" she said pointing at the canvas bag I was carrying.

"Gwen," I said. "Let's stop pretending. Help me."

"With what?"

"I need to work out what Duncan Winterfinch was looking for."

"You want *my* help? Good grief! Out on the island you allowed me to think you were as ready as I was. But apparently not. You shed your skin and turned into a cold fish. In the morning you barely spoke, just wandered silently around. Now you dare to come here? Bang on the door looking like a creature from the bottom of the ocean?"

"I found this shotgun on Haaf Gruney," I said, and held up the bag.

"A *shotgun*?"

"An old side-by-side. I have a feeling that it means something. It was ... rather well hidden."

The gun was lying stripped down on the table. The mechanism was unlike anything I had ever seen before. The wood enveloped the entire underside of the receiver, and formed a slender curve where there would be edged metal on any normal weapon. I held the barrels to the light of the window. They were coated not with rust, but with dust.

Gwen had a rag and was rubbing the grease off the locks. A deep engraving appeared. JOHN DICKSON & SON, EDINBURGH was etched above a cluster of rosettes. I picked up the stock and recognised the acrid smell of the wax coating the wood. The butt had criss-crossed grooves broken by an almost invisible pattern, like a face in shadow behind a grating.

A squirrel hiding its nose in its tail. The handiwork of Einar Hirifjell.

Gwen reached for the grip of the stock, and for a second there were four hands on the wood. I let go, she dug her nail into the layer of wax and began to scrape it off.

What has happened to her? I thought. Something about this shotgun had made her mood shift sharply. I had said nothing about the coffin, even less about the letters or the dress.

Gwen opened a cupboard in which was a Hoover and some cleaning products. She took a metal tin of furniture polish and shoved a wad of cotton inside. But when she tried rubbing it on the stock, the cotton just got stuck and was torn into long threads, like ski wax in hot weather.

"This is what we have to use," I said and took a cloth and a bottle of Fuller's Turpentine. The last time I had felt the solvent burn my nostrils was when I removed the swastika from Bestefar's car. This time it was a pattern emerging instead. The rag soaked it up greedily, I poured on more and kept rubbing.

By the time the stock was clean, I could scarcely believe it was wood. For a moment I had stopped rubbing, fearing that the pattern was painted on. But the harder I polished, the more became visible. It looked like a painting whose meaning is individual to each beholder. From a reddish-orange depth, blue and black lines spiralled outwards wildly, like a blazing fire. The pattern changed depending on where the light struck it. It glinted and new shades became visible whichever way I looked. It was like a viper's nest slowly stirring to life after winter. In the centre of the wood there was a dark, craggy concentration, a maelstrom the colour of dried blood, with thin strands swirling around it. I had seen this in the flame-birch, but the scars of this wood carried something deeper and more impenetrable.

Gwen broke the silence. "Exquisite," she said. "Divine! Walnut of the highest quality. The queen's jewellery box could be made of this."

I looked at her out of the corner of my eye.

"You don't have to act so surprised," she said. "All British servants learn to recognise the emblems of the upper class. We polish their furniture and clean their weapons."

Did she think I was stupid? "The shotgun must be ancient," I said. "The manufacturer has probably shut down."

"This has never seen the inside of a factory," she said, running a finger down the barrel. "It's handmade. A sporting gun of the finest pedigree."

"The mechanism looks rather ..." I searched for a word. "*Odd*?"

"*Old* or *odd*? You really haven't been in Old Blighty very long. Age and wear and tear are badges of honour. I would guess that this is seventy to eighty years old. That's nothing for a British crown jewel. Dickson still exists, obviously. I've passed their shop many times. What's the serial number?"

From the trigger guard, a narrow tongue of blued steel ran down the shaft. There were four digits. But the number was not engraved; the metal *around* the numbers was meticulously shaved away, and the number 5572 appeared raised. I mumbled it to myself.

"You ought to go to the gunmaker and ask them about the history of the weapon," she said. "You were going to go to Edinburgh anyway."

"Sure, but what could they tell us? It's just a shotgun."

"*Just a shotgun?* Everything as old as this has a story. Especially everything British, handmade and worth a fortune. They should be able to help you find out where this wood comes from. And how such an expensive gun fell into the hands of a coffin maker. Have you been to Edinburgh before?"

"Never."

"To any big city?"

"Just Lerwick."

"Now why does that not surprise me, Edward?" She shook her head. "In that case you should leave the car in Lerwick and go by ferry to Aberdeen. Take the train the rest of the way."

She looked at her scratched-up wristwatch. "The ferry leaves in five hours. You'll make it."

"Why are you phoning *here*?" her father asked.

" I was just wondering if Hanne was around."

"Are you pulling my leg?" he said, and hung up.

The coins rattled through the telephone box and landed in the little metal drawer. I counted the cost of a failed attempt.

An attempt which had got dangerously out of hand. Because even down the crackling telephone line I detected the surprising emphasis on "*here*". Through the smeared pane of the telephone box I stood watching the sheep gathering on the slope. I caught a glimpse of the sea between two hills. A fishing boat drifted into view and was gone long before I managed to digest what Hanne must have done.

I felt both excited and nauseous as my fingers ran over the dial. A telephone number I knew better than any other, which all the same felt foreign.

I had never had to call home, to Hirifjell.

I could imagine it ringing at the other end, every single detail stood out; a chest of drawers in an empty house on an empty farm, a telephone next to the photograph of Mamma and Pappa.

There was a click, and I started when the crackling was broken by a "Hello". For many years her voice had offered me promise, but now it filled me with despair.

"You've reached the Hirifjell residence," she said.

"Is that *you*?"

"Haha, yes! It's me!"

"I spoke to your father," I said.

"Was he annoyed?"

"No more than usual."

"Pay no attention, Edvard."

"Listen," I said, "have you just got to the farm?"

"I've been here for four days. Just dropped in to start with. I was quite cross, I have to admit. The potato plants are fine. I checked all the fields, not a speck of blight. I let myself in and had a look around. He left this behind, I thought, he left but he's coming back. The next day this summer weather arrived. And maybe . . . well, we can take this up in bed when you get home, Edvard. I'm sorry I was upset with you when you left. You did the right thing, and I'm proud of you."

Fuck, fuck, *fuck*, I thought and felt the vipers crawling in my stomach.

"Are you still there?"

"Yes. Of course. I . . . I'm just surprised."

"What are you up to?"

"I'm getting these buildings in order. He left them to me."

"Wow! So now we have a holiday home in Shetland?"

I bit my lip. A seven-sided coin dropped into the silence between us.

"Where are you sleeping?" I said.

"Upstairs in the log house. Our old room. Took my books with me. It's lovely to study in this peace and quiet. I can smoke here without anyone nagging me. And I'm trying to be a farmer. I mowed the lawn and weeded the vegetable beds at least. What should I be doing with the strawberries?"

I put my head in my hand.

"Are there a lot?"

"Enough for an army."

"Just leave them."

"But they'll rot, won't they? I was thinking about picking

them and taking them to the old folk's home. Wouldn't that be nice?"

"Listen, Hanne. You don't need to look after the farm. I'm coming back soon. Just leave it as it is."

Another coin dropped.

"What's going on, Edvard?"

"A lot."

"Listen. I . . . I've always thought that you would end up here. Isolated and . . . locked away. Now I've been here and I've felt it myself. I've missed you. I like it that I'm here and you're there, and that you're coming home soon."

I stared at the wet sheep beyond the stone walls and wished that my life was as simple as theirs. But maybe sheep were more complicated than I knew.

"Hanne, it's not a good idea to have . . . expectations."

"I've thought about it a lot," she said. "What you're doing now—"

"Don't say anything else. Listen, Hanne. I can't promise I'll be the same when I get home."

"What do you mean, not the same?"

"I'll call in a few days," I said. "I'm out of coins." Her voice was cut off and replaced by a short beep. I stepped outside, still with a fistful of twenty-pence coins.

The ferry set off from Holmsgarth, turned and passed Lerwick. The city fell into place in front of me: the thick smoke rising from the tall chimneys, the white, crossed cornerstones on the brick houses, the fishermen on the dock and the old cannon at Fort Charlotte.

Fifteen minutes later, I saw the craggy coastline drift past. Monotone, deserted. Black cliffs, everything else washed away by the sea.

A girl came out on the deck. She leaned over the railing a few metres away from me. She was wearing a tweed jacket that accentuated her backside.

I did not move, pretended I was not surprised, and we stood like that for a minute. Then two.

Suddenly, as if on a signal – or perhaps our attraction had developed simultaneously – we moved closer to each other. Her arm grazed mine.

"Are you going to Aberdeen?" I said.

"I have a little errand to take care of, yes. And you? Have you treated yourself to a cabin?"

"I took one of the cheap deck chairs."

"And the shotgun?"

"It's in the car."

"So you ignored sound advice and took the car anyway?"

I suppressed a sheepish smile. "What's that place over there?" I said.

"Must be Troswick. Why do you ask?"

"I just like speaking English. I need to improve. So I don't sound like a foreign doctor."

A breeze swept past us. She turned up her collar and looked at me.

"Would you like me to help you improve?"

"I think we're well on our way," I said.

"Just one thing: Were you planning to march into Dickson's – like that?"

"What do you mean, 'like that'?"

"You can't – simple as that. It's . . . oh, never mind."

We passed the southern tip of the island. The breakers against the distant headland looked smaller and smaller.

"When you booked the deck chair," she said, "was it because you knew someone had booked a cabin?"

"Not at all," I said.

"Well, I'm going below," Gwen said when we had left the Shetland Islands behind. "Are you coming?"

"Take off your socks first," she said, locking the cabin door. "Then you can take off your trousers."

220

She turned her back to me as she unbuttoned her coat. A black jersey hugged her shoulder blades. I could see the straps of her bra and a vein throbbing on one side of her neck.

"Are you saying what I think you're saying?" I said.

"Absolutely. Socks off. Then the rest. All the way down. I'll do the same, but with the light off."

"I don't know if I can accept this gift, Gwen."

"Come on. This is your English lesson."

"And what does a pair of socks have to do with my English?" My throat was dry and the blood was tingling in my forehead, but the proof of my body's urgent desire to impose its will was throbbing in my trousers.

This is the price you pay for the masked ball, I thought. Anyone but the two of us, the *real* Gwen and the *real* me, would have slept together a long time ago. This is what we do to be able to keep our masks on. I had to sleep with the pilot to get past the reefs.

That's a flimsy explanation, I told myself. The fact of the matter is that you fancy the pilot and everything that comes with her.

"Listen," Gwen said, placing her index finger on my chest. "You asked me to teach you English. I took that to mean that you had made up your mind. English men are white as sheets. They always wear black socks. There's nothing more unattractive than a pale naked man in black socks. That's why you always take your socks off first. Come on. Be a good Englishman."

I woke up and looked out the cabin window, felt the ship moving through the greyish-black sea. Hanne's body still lived in my fingertips. The curve of her bottom, the tightness of her thighs, the joints of her spine as I ran my finger down her back. Gwen was different, rougher and broader, and when she straddled me in the middle of the night, my hands did not find what they were looking for, it was as if they were trying to transform what they felt into Hanne. At the same time there was a part of me that whispered, *Yes, go for it, now that you've started, you might as-well go all the way.*

"What was that you were saying last night?" I said to Gwen when she was awake.

She pulled away the duvet. Chuckled. Reached for her suitcase and took out a pack of Craven A.

"Just before I felt the sea shudder beneath me?" she said as she lit a cigarette.

"Around then, yes."

"That was part of your English lesson."

"Now I'm nervous."

"It was Queen Victoria's advice to women on their wedding night. 'Close your eyes and think of England!'"

"But you didn't just *say* it," I said. "You shouted so loudly they heard it in the engine room."

"Good fun," she said, passed me her cigarette and twisted out of the bed, dragging the duvet with her. The cabin was not very large, and I watched her as she walked the few steps to the shower. Her shoulder blades were bare, and her sizable bum stuck out from beneath the duvet.

I dozed. It was still dark outside.

She emerged fully dressed, in a white blouse and a black skirt.

"It's almost five o'clock," she said. "We have to be quick."

"I thought the ferry didn't dock for another three hours?"

"Your English lessons aren't over. You'll have to get changed before we go to Dickson & Son."

"Sorry?"

"You'll have to get changed."

"But it's just a gunmaker. Anyway, I don't have much to change into. Other than clean underwear. I washed them in a pot of boiling water on Haaf Gruney."

"Oh, my!" she said. "I suspected as much." She ploughed her way through the clothes scattered on the floor and reached for a worn leather suitcase that stood in the corner. "Listen. A gunmaker is not just a place that 'sells guns'. You will be standing at the very altar of British refinement. Where old money is burned as if it was Polish zloty. Where the rich can justify their purchases with the excuse that it will go on to the next generation, and the one after that. There are no, I mean absolutely *no* limits. The prices of these weapons are astronomical. And you want to march in wearing a scruffy anorak and trainers with holes in them? Is that really how things work in Norway?"

"I'd say most of Norway dresses better than I do."

"They would think you'd stolen the weapon, that you're stupid enough to see how much you can get for it. They might even call the police."

"And you're telling me this *now*?" I said, bolt upright in the bunk.

"The staff will be lovely, but they hate time-wasters and nit-wits who lower the standard of the shop, and they hate wannabes even more, and by God, you can be sure they'll make you under-stand that you don't fit in. One look from them will bring you to your knees. Your only hope is to make them actually believe that someone in your immediate family once bought a Dickson. You have to be presentable, or you're nobody."

She had changed completely. All of a sudden she seemed genuine, through and through. The awkward expression some-where between aloof and mysterious had been replaced with a lively and insistent eagerness. Was this the same Gwen Leask? Could a girl who babbled like that actually lie?

"How do you know all this?" I said.

"The current Winterfinch family is seventh-generation old money. But we don't have time for that now. And anyway, you know perfectly well I can't talk about my employer. The fact of the matter is that you look hopeless. Utterly unpresentable. Your haircut is cool, I'll give you that. Passé, but wonderfully eccentric."

She undid the leather straps on the suitcase, pulled out two pairs of corduroy trousers and weighed up the difference in the brown tones. She unfolded a tweed jacket of a deep-beige, subtly interwoven with criss-crosses of green and dark-red threads to form tiny, almost invisible squares.

She put her hands on her hips. "The point is that you need *worn* clothes. You don't want to march in there wearing something starched, straight from the shop. Now hop in the shower, and don't forget to shave. Preferably with the door open so I can have a peek. Chop chop."

"Where did you get these clothes?" I said, changing the blade on my razor.

"A small loan from Quercus Hall. Highly inappropriate, but never you mind about that."

I dried off and was handed a white shirt. "Egyptian cotton," she said. The fabric was tightly woven, but soft and pliant. Warm against the skin, and not a single wrinkle. The trousers were a little loose, but she produced a leather belt with a golden-brown sheen and slung it around me to measure the length.

She stared at my trainers in disgust. They lay where I had kicked them off, the laces undone. All my life I had worn trainers or safety boots from the agricultural cooperative in Saksum.

"Stiff shiny shoes would give you away at once. You're a ten and a half, right?"

I turned so she would not see my face.

It pained me, the vision of Hanne so pure and innocent in the garden back home. The two of us. From the first time I saw her moped headlamp by the postbox when she was fourteen, until the day she stood in Alma's bridal dress. Her grey dress in the church.

It was too late. I had resisted the jump without realising I had already taken it.

From a dark-blue cloth bag Gwen took a pair of brown walking shoes. The shiny patina concealed a number of nicks and tiny scratches, like the back of a sunburned galley slave.

"A pair of John Lobb Derbys, with no adornment. Show me how you tie them."

I bent down, tied a bow and stood up.

"Oh dear," she said, getting down on one knee. "These aren't trainers."

She adjusted the lengths of the laces and folded the ends in two loops. "Repeat after me," she said. "Turquoise turtle knot. The only way to tie shoelaces. Watch."

I looked down at the top of her head, which bobbed in time as she tied the knot and muttered at the shoes.

"The loose ends have to be turned *towards* each other. That's important. Then you cross them . . . like this, and then you take a new loop, like *this*, afterwards you let go of this end, hold it with your thumb, like *this*, then you let go of this end, hold it with your thumb, like *that*, then you grab it with your index finger, *there* . . . then you slip it around the loop and pull it out again and tighten it underneath the first loop. Look at that!"

She worked on me as though I were an unfinished statue. Clipped a hair from my eyebrow with a tiny pair of scissors. Conjured up a vial of Truefitt & Hill Sandalwood and patted two drops on my neck. Found a loose thread on the jacket, clipped it off, adjusted the collar, reconstructed me.

"This is fun," I said, and I meant it. "Where did you learn all this?"

"Don't ask so many questions. How were you planning to transport the shotgun? Dearie me."

"Well, in the grey bag I showed you."

"Through the streets of Edinburgh with the barrels sticking out? Oh Lord."

"The barrels don't stick out. I am not—"

"Anything but *this* would be unforgivable," she said, reaching under the bed and pulling out a slim suitcase. The corners had tarnished brass fittings, the leather was scratched, and one end was covered with faded luggage labels from train stations around the colonies. "This will be perfect," she said. "I borrowed it from the rather large collection of sports equipment in the main building."

*

I wished there were one hundred and twenty Norwegian miles between Aberdeen and Edinburgh, but in reality it was closer to twenty. In my head I had transposed English miles with Norwegian, and imagined travelling more than six times the distance. Driving in the low sun, "The Cutter and The Clan" in the cassette player, sharing a pack of Craven As.

Sweet deceit on a dual carriageway. How quickly I got used to it. How much better it was than the asphalt back home. How easy it was to say that this game of roulette was necessary because it served a greater good. How easily regret disappeared into the exhaust behind us.

I had never liked "nice" clothes; mostly they were for occasions I could not bear. Days on which I had to dress up never turned out well.

But these?

The shirt followed the contours of my body, the jacket did the same. I felt like a Leica fitting perfectly in its case.

"Don't mention where you found the weapon," she said. "It is not done to speak of such things in these circumstances."

She had changed too. She wore a freshly ironed, soft blouse and held a thin, ribbed sweater on her lap. I suspected her grey knee-length skirt had been tailor-made and kept in a large wardrobe in Quercus Hall. If Hanne had worn clothes like that, she would look like the elegant daughter of a shipowner.

The only thing that dampened the effect was Gwen's old wristwatch with the scratched-up glass.

"What's the story with that watch?" I said.

"Just a small idiosyncratic touch."

"Where did you get all this stuff? Did you study fashion design or something?"

"Oh, no. I know a lot about how to dress up well, but not much else. I'm a terrible economics student."

The road signs counted down the distance to Edinburgh. Forty miles. Thirty-two. When it got down to eighteen, she asked me to pull in to a lay-by.

"Just leave the motor running and hop outside with me. You smell too clean, a little too . . . new. You're missing a touch of nonchalance. Stand in front of the exhaust pipe. No, there. Turn around. Here, take this pipe. Dunhill. Here's a classic tobacco. Three Nuns. Yes, pop off the lid. There should be a lighter in your pocket. There, yes. Let the smoke settle in your clothes. Mmm, exactly. They'll recognise it. A smell to match the aftershave. It's faded a little now. In fact, it's faded a little too much. You need another drop. That's it, yes. And the other side of the jaw. Keep smoking. Blow on the jacket, don't spill the ashes. *There*. Now the scent is just noticeable, and only when you come right up close to someone. Let's wrinkle your shirt a little. No, let me do it. Just a little. There you go. The clothes still look nice, you can see they cost a fortune. The point is that you don't care. You've just spent a day outside with motoring and other sports. A hint of carelessness shows that money's not important to you. You should never squander, never destroy anything, and for God's sake *never* show that you're rich — just live hard, have fun and wear things out."

She got back into the car and spread the roadmap over her knees. "Now, let's drive."

"Remarkable."

The two men mumbled to each other, pointing out the proof-marks under the barrels, commenting on little details about the mechanism. They were both in their late fifties, two gunsmiths wearing bluish-grey aprons stained with spots of oil. The man who had introduced himself as Mr Stewart was rubbing the wood with a white cloth. They held the barrel and stock between them, ran a measuring tape from the end of the stock to the trigger and noted the dimensions down to a sixteenth of an inch.

"Dickson Round Action, bar in wood," Mr Stewart said. "Our pride and joy. Exceptionally rare."

"How old is it?" Gwen said.

"Pre-war, certainly," said Mr Stewart.

Gwen did not seem surprised. "Thirties?" she suggested.

Mr Stewart smiled, pleased that he had played up his point. "No, miss. The *Boer* War. It's from 1898. Exceptional woodwork, to say the least. Are you after a valuation?"

"More than anything we would really like to learn about its provenance," she said. "And be sure that it's safe to use. It belonged to a distant relative and we're not absolutely certain of its condition."

I could not grasp what she had actually done, other than demonstrate rock-solid self-confidence. The entire shop was now at our beck and call. I stood in the background, studying her and the shotgun, which she had handled not like a weapon, but as if it were an interesting antique, and the gentlemen behind the counter met every question with a well-informed, polite response.

The interior of John Dickson & Son looked like a snapshot from a safari lodge. The shop was small, but old money oozed from its seams. It was neither an office nor a workshop, or even a shop, more like a drawing room and with some of the discretion of Landstad's funeral parlour, where the presence of a cash register would be an affront. Near the entrance there was a selection of woollen clothes in earthen colours. The tags made me wonder whether the prices were in Norwegian kroner rather than pounds.

"Just a moment," Mr Stewart said, disappearing into the back room. He returned carrying a large, leather-bound book with "1893–1905" imprinted on its spine. It smacked heavily against the table. He turned to a page and then went to fetch four other books. Gwen inched closer to get a better look and nodded now and then as they pored over illustration after illustration. Her knee-length skirt and ribbed woollen sports sweater were completely at home here. She was the vase on the table, and she knew it.

Around us were five large vitrines, each containing rows of shotguns with discreet engravings. Price tags were attached to the trigger guards with twine. One of them cost more than I had paid for my car and looked rather simple and austere. The gun next to it was twice as expensive. Another cabinet held new weapons. These had no prices attached.

"The books contain the specifications of every weapon produced here since 1820," Mr Stewart said. "We use the serial number in the index volume to find the information in the other volumes. One book for the weapons as they were on delivery, another for repairs. Each book has one page reserved for each gun. If the page is full, we refer to a new book. This one here ... let me see." He ran his fingers down the lines of brown ink. "Here it is. Number 5572. It was made to measure for one Lord Ingram of the Isle of Skye. He took possession of the gun on August 24, 1898. ¼ and ½ choke – quite common. Top of the range back then. 150 guineas."

"What do they cost now?" I said, and was immediately embarrassed by the ensuing pause. Gwen squirmed.

"A*hem*," said Mr Stewart. "They start at thirty thousand guineas. Depends on the wood and the engravings." He said it quietly and casually, as though I had asked the way to the loo. "Well, we're fortunate to have all the books intact. I see that we accepted it in part exchange in 1922. Then it remained unsold for a long period. Very typical, unfortunately. You can always study the history of war through the life of a British sporting gun. Many of the owners were officers who never returned for their weapons. Not to mention all the craftsmen who fell. The First World War was the darkest hour for gunmakers in this country. Indeed, this weapon remained unsold for nine years. It was not until 1931 that it was sold to ... let me see ... is this an H or an M? Yes, a Major General Mortimer. He must have had an entirely different build to Lord Ingram, as the stock was shortened by one and three-sixteenths of an inch and turned two and a quarter degrees inwards. I would estimate that he was five foot six inches tall and had a strong jaw. The weapon seemed to be in regular use; we had it in for servicing every other year until 1940. After that it disappeared from our records."

"Another war," I said.

"Precisely," said Mr Stewart. He turned the pages back and forth, jotted down codes and page numbers, fetched new leather-bound volumes and searched through their handwritten pages.

My pronunciation must have been passable. The strange thing was that I did not feel overdressed. Walking between the stone buildings and down the busy streets of Edinburgh, with so much traffic that the only thing I could hear was a constant droning, I thought the clothes made me fit in, made me rather enjoy the person I had become.

"Here it is," Mr Stewart said and pointed at a page. "In the summer of 1954 we polished out rust in the barrels and gave the weapon a general overhaul. Very typical at the time. Neglect when an owner passes away. The new owner was called Westley, lived in Lerwick on Shetland."

Gwen looked at me, a look that said *We're getting close.*

"Strange," said Mr Stewart. "In 1972 we gave Mr Westley an estimate for a significant repair. The weapon had obviously been damaged, perhaps fallen down a steep slope. The stock was cracked, the fore-end badly damaged. But it seems the repair was not carried out. The cost of doing such work is rather high, since wood of this kind is particularly complicated to fit. It requires a stockmaker with a master craftsman certificate and at least ten years' experience. But now we come across something interesting. In 1898 the weapon was delivered with the stock graded as 'deluxe nr 4'. That is, well above standard. But this wood is far, far more costly than that. In fact, I'm not sure I've ever seen anything like it."

He held the stock in both hands, turning it slowly so that the light from the windows brought out more and more nuances in the walnut, and then he slipped into a world of his own. "Hmm," he said after a minute or two.

"How many classifications of walnut are there," I said.

"Ten standard grades," he said absently. "Then there are also the special classes, including specific categories for exhibition weapons. Exceptional colour and sheen, but too fragile to be used as a working weapon. Then there is 'Circassian Grade' – just as beautiful as the exhibition classes, but straighter and more sturdy. Named after the Circassian women, known to be the most beautiful women in the world."

Reluctantly, Mr Stewart put down the stock. "The wood alone makes this weapon very valuable. *Very*. At the right auction a stock blank like this would have obtained a price corresponding to a brand-new Jaguar. This is probably from the root of a walnut tree that was at least four hundred years old. Impossible to get hold of something like this these days."

He ran his sleeve over the stock and rubbed a new sheen in the wood. "Is it for sale?"

"Possibly," said Gwen.

"Both the weapon and a stock blank like this would be, hmm, attractive to us, for a discreet transaction. Its only shortcoming," he said, running his finger over the criss-crossed checkering, "is that the stock was not fashioned by a master stockmaker. Oh yes, he was exceedingly skilled. The workmanship exhibited here is rare. He was undoubtedly valued at the time, but he would have had to acquire experience over many years until he was in the foremost class. And that he would have been. There is the minutest level of uncertainty visible in the detail. See here – a tiny curve where the checkering should be straight, an infinitesimal splinter in the hollow of the tongue and the trigger guard. Clever nonetheless. Only five or six stockmakers in the country are worthy of such a blank, and even they would have lost sleep over flaws."

"But is it poorly made?" I said, and stopped myself from blurting out that it probably was the first time Einar had made a weapon stock.

"Poorly made? Sir, it is near perfection. Let's say that it delivers ninety-nine per cent instead of one hundred. He has even got the *feeling*."

My blank look made it clear that I did not understand.

"With a thoroughbred shotgun," he said, "the grip should be slender. The thickness of a lady's wrist. But it's one thing getting the dimensions. Two weapons can look identical and only one will come alive in your hands. The grip should also *feel* as though you're holding the wrist of a lady. That's what this one does."

He picked up the weapon and let the wood shine under the lamplight. "There is something . . . *different* about this wood – something I can't quite place. It reminds me of a story Mr Battenhill told me, something about a consignment of French walnut—"

"Battenhill?" Gwen said.

"Our oldest stockmaker. Buyer of walnut for more than sixty years. Retired now."

"That's a shame," Gwen said.

"He still comes here every day. I'm certain he would be happy to speak to you. But I must warn you, he can seem somewhat . . . *brusque.*"

"Oh?" I said.

"He's ninety-three years old. And for eighty of those years he didn't use ear protectors."

The old stockmaker was due to visit on this day too. Gwen and I were able to wait for him in the office, and we barely exchanged a sensible word; it was as if we were awaiting a verdict.

Before long we heard a booming voice outside and Battenhill appeared, an imposing old man, not least because he spoke loud enough to be able to hear himself. But when he caught sight of the weapon, he lowered his voice and whispered: "Good heavens, it's one of *them.*"

He picked up the shotgun as though it were a lost child and said that it had been twenty-six years since he last saw a weapon with wood like it, a double-barrelled African rifle that had once belonged to Ambassador Cleve. It went to auction, and one of the descendants of General Haig, the British Commander-in-Chief during the First World War, had offered an outrageous sum for it.

"But what's the connection?" Gwen said.

"The wood carries the wounds of war," Battenhill said and showed us a darker area on the shaft.

He rummaged through a drawer and brought out a small glass bottle. The smell of linseed oil spread through the room as he rubbed it into the wood. The pattern grew more and more intricate, as though he were performing alchemy.

"It's a little dry," Battenhill muttered. "Look there," he said reverently. "One moment the pattern is symmetrical, like the outspread wings of a pheasant. When I rub it, like this, the light hits it from a different angle and brings out the deeper layers of the wood. Every quarter of an inch has taken many years to grow. We're looking back through the centuries. But the year that is clearest of all is 1916."

He continued to massage in the oil, and with every moment the pattern became clearer. He began to tell us about the Scottish regiment, the Black Watch, of their advance during the Battle of the Somme, and before long it became clear that the story was also about me: the year my history began was far, far earlier than the year I was born; *my* history began four hundred years ago, when sixteen walnut trees sprouted near the Somme.

11

AT FIVE O'CLOCK ONE MORNING IN SEPTEMBER 1916, the old stockmaker began, a company from the Black Watch was lying in the trenches, waiting for the signal to attack. When the bagpipes sounded, they advanced together with the Cameron Highlanders into the remnants of what had once been a lush forest. Artillery had levelled the area into a mash of mud, charred stumps and battered corpses. Like the soldiers who had been in the previous attack, they wore kilts. The corpses they advanced through were unrecognisable, but they could see where the individual units had fallen, their various tartan colours still fluttering amongst the carnage.

The Battle of the Somme had been raging since July 1. On the first day alone there had been fifty-seven thousand British casualties, twenty thousand of whom had died instantly. Machine guns ran the entire length of the front, and masses died every minute in the most foolhardy advances. Hundreds of soldiers were left hanging in the barbed wire, and it was impossible to remove them. Their flesh rotted in the summer heat until it hung loosely. Even when the bodies could be buried, it was impossible to keep them in the earth because as soon as the counterattacks were launched, they were blown back to the surface.

"It was as though the great powers had mustered a gigantic European collective effort," Battenhill said. The very best machinery, the cleverest engineers, the most employable generation – millions stood along a front line that carved Europe in two; enough manpower and skill to build a pyramid every day.

Fields, churchyards, forests and villages were transformed into infinite mires. Civilian miners dug kilometre-long tunnels and placed so many explosives that the detonations could be heard in England and left craters the size of meteor strikes. Soldiers who had only recently arrived wondered about the small clouds that appeared to be suspended over the fields, which were visible from fifty metres away. Only when they recognised the surreally loud buzzing did they understand these were in fact swarms of flies feeding on the corpses.

The artillery thundered day and night on both sides. The front lines were bombarded with more than a billion tonnes of explosives, but the factories had to manufacture them so quickly that their quality was reduced. Around one in four shells struck the ground without exploding and were soon ground into the earth along with the dead.

But of all the clashes during the Battle of the Somme, the soldiers considered the battles around the woods to be the most barbaric. Splinters flew about like spears when shells struck the tree trunks. The soldiers were forced to attack in tight formation, making them easy targets when the artillery and the machine guns zeroed in. The stumps and roots of the trees made it impossible to dig in and find cover, and the hand to hand combat was so intense that soldiers could be killed by bone splinters from fellow soldiers who had been hit.

The objective of the soldiers from the Black Watch on that morning was to recapture a small wood north of Authuille, on the hillsides facing the Ancre River. This was the seventeenth wave of attacks without a clear victor, and they knew that their opponents were under orders to fight to the death.

The woods were no larger than thirty hectares, but they had been bombarded more intensely than any other place in the Somme. At the most frenetic point in the fighting, seven artillery shells were fired there every second. In the rare intermissions between the cannon salvos, the cries of hundreds of wounded soldiers could be heard from the forest. An entire day might pass before anyone could be pulled out, and by then most were dead.

Forest warfare was nothing new to the Scots. The supreme commanders seemed especially eager to deploy them in these battles, and the Scots suffered huge losses in Devil's Wood and High Wood.

Within an hour of their initial advance, eighty per cent of the soldiers had fallen. Those still alive lost almost all sense of reality, and simply fought their way forward, yard by yard, with bayonets and shovels past the scorched tree trunks and piles of bodies, enveloped by the noise of war. Their supply lines had long since been cut, and the wounded had no hope of being saved.

But with the help of mortars and other high-trajectory weapons, a few soldiers managed to blow up an enemy machine-gun nest and break the German line of defence, after which they stormed a small grove where a few huge trees still stood. They were ancient, thick walnut trees, and even though the branches had been blown off, they were so sturdy that they formed a shield against the artillery. At length the soldiers managed to set up a grouping of machine guns behind the large trees, established good sight lines and eliminated several hundred enemy soldiers.

When night fell, the full fury of the enemy bombardment was unleashed. The tops of the trees burned, but did not collapse. Reinforcements arrived with food and ammunition, and they dug in to brace themselves for the dawn counterattack. They packed earth around dead horses and soldiers, with both hooves and arms sticking out of the barricades.

Through their binoculars the Germans could see that the Scots had dug in, and now that they were sure none of their own men were in the vicinity, they decided to use all means necessary.

At dawn they launched a large-scale bombardment of poisonous gas. The soldiers were familiar with the effects of chlorine and cyanide, not to mention mustard gas, which first blinded and then induced four weeks of hideous suffering as the body slowly surrendered, the intestines rotting while the soldier was still alive.

But this gas was something new. Either it was a devilish, experimental gas so complex that the chemists could never make it to the same recipe, or there was a manufacturing flaw that no-one then, nor at any time thereafter, knew how to replicate.

The shells exploded against the walnut trees, and a heavy, bright-green cloud poured out and settled on the ground. Those soldiers with working gas masks put them on at once, the others had to resort to the old field trick of urinating on cloth rags and breathing through them. But before long they all realised that nothing would offer them protection.

The soldiers grew unruly. Some began to fight one another, others stood up and were immediately shot down by snipers. Those closest to the gas roamed aimlessly between the tree trunks before they passed out and fell into muddy shell holes. The company was bombarded for half an hour while the green gas rained down from the walnut trees. A young captain ordered his soldiers to advance and position their machine guns *in front* of the trees. It was an absurd order, but they set to it and were all shot instantly. The captain himself was struck by a shell and was later found far from his original position.

But the trees remained. The wind began to pick up and carried the gas away. The Scots just managed to get reinforcements in before a counterattack was initiated. They took up positions behind their dead comrades' guns, and laid down a dense, flat line of fire that mowed down hundreds of the enemy in a matter of minutes, and soon the rest could advance and secure their positions. The victory was decisive. In six weeks, fifty-five thousand German and British soldiers had sacrificed their lives in the battles for this small wood.

As Battenhill spoke, Gwen and I had imperceptibly moved closer to each other, until our arms were touching and I could feel her shoulder twitching slightly. But now she pulled away a little and I could not tell if her twitching had intensified or stopped altogether.

"By November 1916 the Battle of the Somme was over," the old stockmaker said, and he then described laconically how the number of losses on both sides stood at 1.2 million dead or wounded. The Allies had gained nine kilometres of terrain, land that was now a sludge of body parts and twisted metal. Few of the soldiers of the Black Watch were given a burial after the battles in the grove, but they had their names engraved on the piers of Thiepval, the memorial to those who were never found or were too mutilated to be identified. After the German capitulation the bodies of eight thousand soldiers remained in the small woods, tangled in mud and roots and unexploded shells.

But in front of the old machine-gun positions, sixteen walnut trees still stood. The canopies were destroyed, the bark shredded and the branches burned off, but they stood. Since everything around them had been swept away, the cluster of trees was visible from a great distance. In the spring, small twigs sprouted and produced green leaves. Along with some orange poppies, they were the only sign of life along the old front, and amongst the British soldiers the trees became known as "the Sixteen Trees of the Somme".

When the authorities established their comprehensive regeneration plan, the so-called *reconstitution des régions dévastées*, the owner of the woods wanted to have the area cleared so he could replant trees there. But this was a task they had soon to abandon; the concentration of shells was so great that the sappers were instructed to make one safe path through the woods. The owner of the woods received no compensation other than free barbed wire with which to fence the area.

Bushes and small trees began to grow, but the field was impossible to cultivate. The owner occasionally walked along his woodland path. From a distance he observed the sixteen charred trees, which until before the war had produced an abundant supply of nuts. Now he could see that the fruit was no longer fresh; the nuts were black and shrivelled, and the foliage soon took on a greyish tinge.

In the meantime a Scottish timber merchant had heard about the trees from a returned soldier, and made the journey to Authuille in the twenties along with a small team of sappers. He made safe a passage to one of the walnut trees and managed to fell it at the root.

When the trunk was transported to a sawmill, they discovered that the wood had an unusually intense golden-red pattern. The merchant put this down to a reaction with the unknown poisonous gas. He got his best people to saw the root, and straight away they recognised that the wood equalled or surpassed the highest-quality grade for old walnut, making it worth a fortune. A fortune that was automatically multiplied by the history attached to the tree.

Walnut itself was a cherished commodity for the British upper classes, which had admired beautiful woodwork for generations. The timber merchant decided that in this instance the walnut should *only* be offered as stock blanks for high-quality sporting guns, at once an ornament and a war memorial, and one that contained both the country's sufferings and its victories.

The owner of the woods agreed to allow the merchant to fell the trees and excavate the roots, but what he received in return is not known. The merchant then realised that the twisting pattern was still developing; the trees were slowly dying, and he wanted to wait before felling them, anticipating that the formations would not reach the height of their beauty until the trees were close to giving up the fight.

From the first tree he managed to carve out twenty-four stock blanks, twelve of which were sold to the country's foremost gunmakers: three to Purdey, three to Dickson, two to Holland & Holland, and four to Boss & Co. The wealthy classes of the war-torn country embraced the story of the wood, and despite the overall economy being weak, the weapons were sold at astronomical prices.

During the thirties, the merchant gradually released the remaining blanks onto the market in order to keep the story alive, and he was careful to draw attention to the part they had played in the war. In 1937, gunmakers received a tip-off that the trees would be ready for harvesting around 1943, and would be available at auction the following year. Expectations were huge; the wood was mentioned several times in *The Field* and described with a single adjective in *The British Shotgun* – "unparalleled".

But then came another war, and when it was over, rumours began to circulate that the walnut had disappeared. Some said that the woods had been burned down in the Allied advance, others that the trees had been cut down and destroyed. But the timber merchant insisted that the shipment was merely in transit, and that all orders would soon be fulfilled.

In 1949, two stock blanks were offered at a firearms auction at Bonhams in London. There was no doubt that they belonged to the missing shipment of walnut – the pattern was unmistakeable and far surpassed the few samples sold before the war. The seller was anonymous, and the merchant sued the auction house. But since the wood had neither stamps nor serial numbers, he was unable to prove that the wood had been stolen.

Another blank came onto the market in 1955, the pattern just as extraordinary as the previous had been, and there was a lengthy bidding war lasting more than an hour. The merchant had by then spent vast sums on lawyers, and stubbornly insisted that he would soon deliver on all orders. But no more samples were put on the market. Nonetheless the shipment – which the merchant claimed would consist of almost three hundred stock blanks – would have been one of the most valuable consignments of wood in history.

The blanks were known by various names. Some referred to them as "the Walnut from the Sixteen Trees of the Somme", or the "Somme Walnut" in memory of the fallen soldiers. But among stockmakers, the wood was referred to as "Daireaux Walnut", in honour of the original name of the forest and the family who had owned it.

12

"WHY AREN'T YOU EATING," SHE SAID.

On the table was an untouched jug of water and two glistening burgers with wooden skewers piercing the crisp, golden buns. The pale yellow cheese had melted onto the raw onion and coarsely minced meat.

"Because of the grip on the shotgun," I said.

"What?"

I reached my arm across the table and grabbed her wrist.

"The gunsmith talked about the grip of the stock on a side-by-side shotgun. It should feel like the wrist of a lady."

"And?"

"You're no servant," I said, squeezing her wrist as I had in bed on the ferry. "You're the granddaughter of Duncan Winterfinch."

There was a long and awkward silence. "I knew that you knew," she said. "But I'm glad you didn't say anything until now. That you let me be ... Gwen Leask."

I shifted in my chair, moved my foot so that it touched the weapon case on the floor. If I had sold the shotgun, I thought, it could be full of money.

"How did you come up with Leask?"

"Oh, it's a common name in Shetland," she said and twirled her fork slowly in her hand. "John Leask runs a removal company in Lerwick. Their lorry passed by just before you asked my name."

"So you've been lying to me all this time?"

"Oh, please," she said, wiping her mouth. "I didn't know about

241

the walnut. Believe me. I knew there was *something* about Haaf Gruney, but I didn't know what."

"Come on! I don't believe that at all."

She behaved as she had at the Raba, concentrated on chewing when she wanted to hide her expression.

"Nonetheless it is the truth. Grandfather had his secrets, I knew he was haunted by something. The island was ours but he was evasive when I asked why he let that strange man live out there, a man who, according to rumour, was a murderer. He was angry with Einar, but until today I had no idea why. It's because Einar kept the walnut for himself."

"I wouldn't be so sure about that. Why would he have decided to live right under your nose?"

Her eyes narrowed. All of a sudden we became the ambassadors for the inherited outrage of two families.

"Could the wood be somewhere on Haaf Gruney?" I said to smooth things over. "He did make the gun stock, after all."

Gwen pushed a gherkin to the side of her plate, cut off a piece of bread. Using her fork, she mopped up the gravy with the bun. I wondered how much she had tidied up the story too, to make it more appetising.

"Impossible," she said. "Once, when Einar was away, Grandfather arranged for a crew to search the entire island. They were at it for four days. Smashed open the locks, dug holes in the earth, even removed the panelling from the walls. I asked him what he had been looking for. His reaction was strange, and evasive. And he wasn't the kind of man you asked twice."

"Why didn't your family clear out the buildings on Haaf Gruney when Einar died?"

She looked at her plate, waved her knife and fork aimlessly before placing them on the table and wiping her mouth. By the time she looked at me again, she had discarded her snootiness.

"Because I . . ." she began. "Oh, dear Lord, this is going to sound so childish, but I've been planning this for years. I was waiting for *you*."

"You were waiting for *me*?"

Gwendolyn reached across the table and placed her hand on mine. "Yes, you – Edward Daireaux Hirifjell."

She told me how her sheltered life had crumbled when her grandfather lost his balance at the top of the stairs, fumbled for the railing, fell down and died. The sheriff gave them a copy of the deed for the freehold, namely Haaf Gruney, on which it stated that a certain Edvard Daireaux Hirifjell, born in 1968 and therefore the same age as her, would inherit that right when Einar Hirifjell died.

Gwen had gone out and looked at the island through a pair of binoculars. There was peat smoke rising from the chimney, and she thought about the strange coffin maker she had occasionally seen near the ferry terminal, but had never spoken to. Did he really have an heir? It was then that she first began to imagine what I might be like, and how I looked.

A few years later, word arrived that the hermit of Haaf Gruney was dead. She had been living in Edinburgh at the time and went up to the funeral hoping to meet me. But the only person she met at Norwick was a quiet old man who drove a Norwegian-registered Mercedes. Gwen told him that the Winterfinch family owned Haaf Gruney, and asked whether I would be making a claim for the freehold. "That will be up to my grandson," Bestefar had answered in broken English, before getting into the car and driving off.

This made Gwen even more curious. She took *Zetland* and stood on the shore of Haaf Gruney for the first time. She asked herself why her grandfather had had the island searched. This was before Agnes Brown had tidied up, so the buildings were still unlocked. Gwen went into the workshop where Einar had made the coffins and began to go through his belongings. But it felt wrong, so she left. A couple of days later she returned, but by then someone had been there and changed the locks, which aroused her curiosity even more.

Her lawyer suggested contacting me in Norway, to clarify

everything. But Gwen thought, No, let's wait. Dispute his right, nothing more. He's related to the man who was hiding something from Grandfather. If I cut all ties to him, the final clue will be lost. Best let the matter stew. Edward Daireaux Hirifjell will come.

"So yes, I admit it," said Gwen. "I hoped that you would show up, slip up and leave some clue for me."

"That is cynical," I said.

"Not really," she said. "I didn't know that your mother and father died in Authuille. But what about *you*? Why didn't *you* reveal your true colours? Why did you shuffle around sheepishly, pretending to lap up every little thing Gwen Leask said? Tell me that, Monsieur Daireaux!"

She pronounced my name in impeccable French, better than I could have, and it felt as though she owned more of my past even than I did. Hearing my family name spoken like that fanned the embers of the biggest question of all, and I sat thinking about the forest my mother's family had owned. I wondered what it looked like now?

"Hello? Nothing to say to that?" she said.

"Yes," I said calmly. "Because I was beginning to prefer spending time with you more than digging up the past."

For a moment she was silent. She stuck her fork into a lonely little potato glistening with fat and speckled with herbs, and put it in her mouth. She cut the gherkin in half and ate that too. I copied her, let the taste of vinegar sit on my tongue for a long time before swallowing.

"I enjoyed it," she said. "Being Gwen Leask. Driving around in that vulgar car of yours, wandering about at night. Maybe I was already in love with you without realising it."

Her eyes had that glimmer of dreaminess, like the moment we first met. Now, I thought, now the sea we're sailing on is beginning to get rather deep.

"After Einar's funeral I went back to Edinburgh," she said, "but I began to spend my holidays in the stone cottage. I liked the thought of a distant soul living in the dreamland of Norway, someone who would be drawn here sooner or later. You became a small, unopened box of chocolates in my life. I had almost forgotten about the whole thing when the shop assistant called me out of the blue. That very evening I saw you rowing across the inlet."

She was different now. More attractive, as though the rigid features of her mask had slipped off.

"I had convinced myself that you were sleazy, that you were trying to swindle me. But you were handsome. Distinguished, in a way. With intentions that clashed with my own. You surprised me. I have never met a man who would drag a dead sheep onto land."

I cleared my throat and thought about my journey home. Gwen Leask and Edvard Hirifjell had travelled to Edinburgh. Gwendolyn Winterfinch and Edouard Daireaux would be making the return trip.

"One question," I said. "What will you do if you find the walnut?"

"That depends on Einar's reasons for hiding it."

She called a waiter, said no to tea and asked for the bill. She glanced at her old watch and calculated the time for the journey home, told me there wouldn't be enough time to drop by her flat.

"What's the hurry?" I said.

"The gunsmith made my grandfather sound like a greedy timber merchant who had accidentally stumbled upon something valuable."

"Well, greedy or not, he—"

"There's one question that has not been asked," she interrupted.

"And that is?"

"Why Captain Duncan Winterfinch was so obsessed with securing the sixteen walnut trees from the place where he and the rest of the Black Watch had fought in 1916."

She was not in bed. I turned over and said "Gwendolyn?" to an empty room.

The cottage was dark. That night had been much more intense, as though we wanted to wrench the innermost secrets from each other. Dig out the lies and see what remained. Part fear and part confusion, released in sweat and indefatigable desire.

Her alarm clock showed a quarter to three. I went into the living room, past the messy coffee table with the music magazines. In the dark of night I saw a light on the top floor of Quercus Hall. We had agreed not to go there until morning.

Back in the bedroom I glanced at the fancy clothes that were strewn across the floor, found my old clothes in the suitcase. I crossed the garden in the pale moonlight. The sea was calm at the cliffs below. The waves rolled in sluggishly, only visible as hazy movements. It was raining again. Not an intense shower, just big, warm drops.

The front door was ajar. Tall and wide, it slowly eased open. The room inside gradually revealed itself as my eyes adjusted to the darkness. A large hall with a double staircase leading to the second floor. But it was not in the grand style I had expected of a manor house. Straight, clean lines and a high ceiling, but economical. Large floor tiles with a compass rose in the centre. Space for probably twenty people in the vestibule. Tall doors of carved wood.

Behind a curtain I found the light switch. A row of globe pendants lit up, suspended from the ceiling by thin metal tubes. Square patches on the walls showed where paintings had once hung. I walked up the stairs. The banister curled like a fat snake of glistening mahogany, the uprights were in the shape of elongated hourglasses. A dark corridor lay ahead of me. I could not find a light switch, but further along I saw a line of light coming from the bottom of a door.

"Gwen," I said.

No answer. The door was locked, so I continued walking along another unlit corridor. I smelled mould and old leather as I walked down another set of stairs and noticed a soft glow coming through a frosted-glass door.

When I opened it I was greeted by fresh air; I realised I had stumbled upon one of the secrets of the house.

An internal garden. Large and lush, and damp with rain. It was so overgrown that it was almost impenetrable. Tall, exotic trees stood like pillars amidst round, broadleaved bushes. I was struck by what I had been missing in Shetland, something Duncan Winterfinch had, but that Einar was denied: the sound of raindrops falling on the foliage of large broadleaves.

"He used to stand up here and watch me," she said.

The voice came from above. I craned my neck and searched for her while drops of water ran down the leaves and fell onto my forehead. I followed the row of windows along the walls that surrounded the garden and spotted her on a veranda that jutted from the first floor. I moved some branches aside and took high steps through the underbrush to a spiral staircase in the corner.

"It's practically a greenhouse," I said.

"No," she said and leaned over the railing. "It's an *arboretum*. A botanical garden for trees. Eighteen different species of wood. Originally there was a glass roof to maintain a warmer climate, but it was shattered in 1933. The shards rained down on the gardener and almost killed him."

She guided me to another spiral staircase that led to the second floor, where an internal veranda ran all the way round. From there we could look into the treetops. The longest branches brushed softly against the railing. It was like being in a tree house in the middle of a forest. The waves from the sea were barely audible. Beyond the walls, the wind could probably get so strong that the salt spray would damage anything that tried to grow. But in here everything was shielded; the walls held the plants in a gentle embrace.

"We have always lived off trees," she said. "Off them and with them. For seven generations the Winterfinch family have been selling timber. From the rainforests, from the Russian taiga, from Norway and Sweden. For decoration, for furniture, for buildings. In our 1901 catalogue, we stocked seventy-eight different types of wood."

She walked to the edge of the veranda, bent her knees a little and bounced up and down to make the floor yield. Not a creak was heard.

"Oak?" I said.

"Yes. Oak has been with us since the beginning, and oak is what made us our fortune. Some say Winterfinch Ltd deforested Britain, but that's nonsense. Only war can strip a country of its trees. And oak especially. The only wood good enough for a warship. For a standard seventy-four cannon ship, thirty-seven hundred oak trees are needed. When H.M.S. *Victory* was completed in 1765, it took fifty-seven hundred fully grown oak trees. The navy's chief administrator of shipping timber was Gregor Winterfinch. He personally inspected every single plank on *Victory*. The documents are in the cellar. One of the signatures belongs to Admiral Nelson."

"I'm beginning to realise that you Brits are different from Norwegians," I said.

"You were too lax in establishing colonies. Gregor founded the family business, and in 1770 he began to import timber from the Baltic Sea coast. Soon he became one of the country's leading timber merchants. From 1858 to 1893 we were the largest of all. Offices in every major port of the empire. We imported everything from ship timber to the most precious woods for making jewellery boxes and walking sticks. Half of all British joinery sold in guineas was made from our wood."

"What is it with guineas?" I said. "The gunmaker used that expression too."

"Mass-produced goods were priced in pounds. But everything made according to the customer's own specifications was paid for in guineas. A standard table – priced in pounds. A bespoke dining table – in guineas. A mass-produced Lee-Enfield – pounds. A Dickson Round Action made to measure – guineas. Racehorses, oil paintings—"

"But what's the difference?"

"Oh, practically nothing. We haven't used guineas in coin or note form since 1816. But a guinea is worth a pound and a shilling. There was a tradition that the master took the pound while the apprentice took the shilling."

A breeze reached the treetops and made the leaves rustle. Some flies were buzzing about.

"Grandfather was not meant to inherit the company, in fact. Stanley was the eldest brother. Grandfather went to the Royal Military College at Sandhurst and imagined an exciting life for himself in the colonies. But Stanley died of malaria after visiting the office in Georgetown. When Grandfather took over, he was already weakened by the war. One of his first great mistakes was to build this house."

"Mistake?"

"They started in 1921 and gave up in 1928. Now come with me," she said impatiently. "Let's go inside."

We walked through three echoing halls, illuminated only by the greyish gleam of the summer night, and entered a windowless corridor that was as long as a schoolhouse. She snorted in irritation that the light switches had "failed again", and we felt our way along until she stopped in front of a door, her movements barely visible in the darkness, like a black cloth folded over another black cloth.

"I idolised my grandfather," she said. "Behind this door is a room that was once the safest place I knew. I always thought he had his reasons for being the way he was. But then you arrived. Forced me to ask myself whether those reasons were quite good enough."

She stretched out her arm. In the dark I saw the luminous hands of her watch form a narrow angle. It was five past three. "You stared at this watch when we met," she said.

"Because it's a man's watch," I said. "I wondered if you were engaged."

"This watch was in Authuille, in the place you disappeared," she said and opened the door. "But fifty-five years earlier."

She entered the dim room. The first thing I noticed was a worn depression in the middle of the floor, like some kind of metal mechanism had spun in circles there.

Then a ghost passed through me, I had a fleeting sense of *déjà vu*, but it disappeared before I could latch on to any memory. It was the *smell* that had roused me. There was a scent of age and furniture polish in the room, but underneath that, like the deep tone of an organ, there was a sharp, earthy fragrance.

"That smell," I said. "What is it?"

She drew the curtains. Ran an index finger along the window-sill and wrinkled her nose at the dust.

"It smells like Grandfather," she said. "Pipe tobacco. Balkan Sobranie Mixture. He smoked in here non-stop for fifty years."

She flipped a switch and a yellow light filled the room. It was a good sixty square metres, and took up an entire corner of the house. One row of windows offered a view of the sea and through another I could see the fields of Unst. I crossed the floor, and from there I could see the veranda and the green foliage of the arboretum.

Gwen stood with her back to the window and observed me silently as I wandered about. In front of a book cabinet stood a massive bureau with worn corners. Small woven baskets on top contained dried-out rubbers, glasses' cases and matte fountain pens. The bookshelves were crammed with yellowed newspaper cuttings and leather-bound books.

A glass-fronted cabinet was filled with whisky bottles. Most of their contents had evaporated. Another cabinet was stacked with flat tins. "This is what you can smell," she said, twisting one open. "Balkan Sobranie."

On the inside of the lid were the words: *A long cool smoke to calm a troubled world*. I put my nose to the dried tobacco. Tried to evoke a hint of a memory, but it did not come. "The jacket I was wearing," I said. "Did that belong to your grandfather?"

She nodded.

"Did you let me borrow it to see the gunsmith's reaction?"

"No. I let you borrow it because you needed a tweed jacket.

Don't be so suspicious. It doesn't become you and I won't have it, not in here."

In front of a sooty fireplace there was a worn living-room set and a low table with an enormous crystal ashtray on top. Above the mantelpiece was a painting of Gwen. It looked like she was in her early teens, in profile and dressed in an old-fashioned checked skirt and sleeveless blouse, looking across to a coastline. I recognised the place, it was a river here on Unst.

"There aren't any other family pictures here," I said.

"Oh yes, there are. Here's his other family," she said and pointed at the wall behind the desk.

The photograph was one metre wide, but no taller than a sheet of A4 paper. I came closer and immediately saw why it was in that format. It was of a military detachment, probably three hundred men divided into six rows. The frame was inlaid with a small brass sign engraved with THE BLACK WATCH 1915.

"He is sitting with the officers in the middle of the front row. The one *without* the moustache. Captain at the age of twenty."

It was a good photograph, with all the faces in sharp grey tones. The soldiers were my age, they seemed cheerful and carefree.

"The Black Watch wore kilts into battle as late as 1940," she said. "The Germans thought they were skirts and called them 'the Ladies from Hell'. Look at the four men standing by the officers – bagpipers. They followed the soldiers to the front line, all the way. It's no coincidence that a corps from the Black Watch played bagpipes at Kennedy's funeral."

"Why were they called that?"

"*Are* called. They've existed since the 1600s. Do you think a Scottish regiment would disband after only a few hundred years? But the name – no-one knows for certain. Grandfather could spend hours explaining all the theories. They wear dark tartan, but my favourite explanation is that they're called the Black Watch because the first soldiers had such dark hearts."

Standing by the photograph, I was struck by that same smell. The smell of tobacco when it settles in a room or in clothes.

Bestefar always came to mind when I caught the smell of cigarillos. Could my father have smoked Balkan Sobranie?

I leaned forward and studied Duncan Winterfinch. The bandoleer cut proudly across his chest, but he was not looking into the camera. Instead he looked to the side, at the privates. A striking number of them were not looking at the camera either, but at their captain. It was impossible to see whether it was doubt or admiration in their eyes.

From a wardrobe Gwendolyn took a khaki uniform jacket. She carefully laid it flat on the glass table in front of the fireplace, the crystal ashtray squeaking as she pushed it away. At first I thought she had not unfolded the jacket completely, until I realised that the entire left arm was missing. The material hung in shreds from the shoulder, the fibres black and resembling rough twine. Blood had soaked into the torn fabric, where it had dried and never been washed out. On the shoulder straps there were three faded captain stars.

"Do you find it unpleasant?" she said.

"No," I said. "Just . . . sad."

She took off the wristwatch and placed it on the table where the left cuff would have been.

"Grandfather had been at the front for almost a year when it happened. He had also fought on the first, the most catastrophic day. And then the Black Watch were sent in to take the woods. What the gunsmith told us chimes with what Grandfather told me. But he left out everything about the trees and the gas. I had never heard about the Sixteen Trees of the Somme until now. I cannot understand why."

It must have been the same gas that killed Mamma and Pappa, I thought. But I said nothing to Gwen, because I could sense that we had each followed our own path during the story, and that we might emerge from it separately.

"All Grandfather told me was that he had been hit by shrapnel. When he came to, he was alone. His soldiers were lying around him, dying or blown to bits. He got to his knees but felt strangely unsteady. It was because his severed arm lay on the ground behind him. His Webley revolver was just beyond his fist. Do you know what he thought was strangest of all?"

I shook my head.

"That his watch kept going, even though it was on a severed arm."

I ran my hand over the cloth. Rough, tightly woven. I had an urge to pick up the uniform, just as I had picked up the dress out on Haaf Gruney, but I stopped myself.

"How old were you when you were given the watch?" I said.

"Ten. Fifteen when he told me the story behind it. How he picked up his own arm, ambled towards the supply lines. In the end he was found next to the Ancre."

Out of a pocket she took a dented, star-shaped regimental badge.

"But he never mentioned the sixteen walnut trees?"

"Never."

"That's really strange. What would have been the harm in telling you?"

"I have no idea. It would have been harmless, even *nice*. But no, not a word. Nor that he had hired *Aainarr* in 1943 to secure the wood."

I remembered what the priest had overheard. *Enough to fill a lorry.* That could correspond to sixteen tree trunks.

"What was he like?" I said. "Around you?"

"The best. He liked it that I always wore his watch. He could be rather crass with other people. He had horrific nightmares, always woke up at three o'clock in the morning. I heard him sometimes, shouting for tea or something. He could be very disagreeable."

In a semicircle on the regimental badge there was something written in Latin. I mouthed it to myself.

Nemo me impune lacessit.

"Their motto," she said. "'No-one attacks me with impunity'."

I stood looking at the uniform. Gwen went to a corner of the

253

room and all of a sudden I heard a loud screeching of metal on wood. I spun around to see Gwen rolling out from behind a folding screen an antiquated wheelchair with a high back. The large wheels were fitted with metal rims and could have come from a horse cart. The material on the backrest was disintegrating.

"He was in this when he came home from the war," she said pushing the chair back and forth with her legs. "In 1921 he was at King's Hospital in London and had fifteen pieces of shrapnel removed. In 1947 he had an operation in the U.S.A. Cost a fortune, but by 1953, he was actually able to walk on his own. It was not until he was in his eighties that he had to use this again on occasion."

Gwen sat in the wheelchair for a long time, then got up, opened a door and stepped onto the veranda. On the other side of the railing there was a primitive lift, a crude wooden box fastened to rusted flywheels, block and pulley and metal wires.

"Of course it would have been more practical if he had simply furnished the ground floor, but this office had the view. Treetops inside, the sea outside."

She pulled a black handle and an electric motor began to hum. We sat on the floor of the wooden box. She pushed a control and in fits and starts we followed the trees down with the smell of foliage, the sea and machine oil in our nostrils.

"Now and again I would take the lift up to see him, dangling my legs over the edge. I'd knock on his door and come from the veranda into the smell of Balkan Sobranie. He would know when I was coming because he saw the lift go down and up again."

"This is *fantastic*," I said.

Gwendolyn Winterfinch smiled.

"Was that not the right word?" I asked.

"Fantastic? Sure. If you want."

"I'll find a better word soon," I said. "I'm working on it."

"It's not so fantastic in the daylight," Gwen said. "I told you that the house was never finished. Two entire wings have no panelling and floors. In 1926, the money began to dry up. My

254

grandfather was actually a romantic. He wanted to make Quercus Hall a tribute to peace, and to all the trees of the world. He loved trees and furniture, became a pacifist after the war. He had no idea how to run a business and his wounds tormented him. What he was *really* interested in was decorative trees for furniture making and weapons. The department supplying building materials, the one that was truly lucrative, was heading straight downhill. Sales dropped off, office after office shut down. It got even worse in 1946, when the situation was similar to that in 1919. No-one had money to spend on beautiful materials; what was needed instead was cheap wooden boards for rebuilding. My mother was born in 1927. In practice, she ran Winterfinch Ltd from the day she turned twenty. Grandfather still ran the trade in exotic trees, but it was like working at British Petroleum and being in charge of sewing-machine oil. So *Mother* controlled everything."

"Why do you say *Mother* like that?"

"My birth didn't fit in with her plans. She'd had enough of children, she wanted to concentrate on the firm. Father just played with his stamps. Vanished from the dinner table like a deer in the woods. I have two brothers who are much older than me. Clear-thinking, efficient men who believe I'm a hopeless, spoiled inconvenience."

"Did you grow up here?" I said. "In this house?"

"To all intents and purposes. My grandparents' mission in life was to bring me up the old way. Clothes, dinner manners and furniture design. I have seen all of Europe's major cities from the back seat of a Bentley Continental. My grandmother's hobby was auctions. When I was twelve, I could estimate the value of any Lalique vase, or date a piece of Miriam Haskell jewellery to the nearest four years. When she died, I was like a walking remnant of pre-war customs."

"But you said that your father taught you to sail?" I said.

She shrugged. "I was bluffing. Grandfather taught me to manoeuvre *Zetland*. Since the age of ten I've been taking it through rough seas with an old one-armed man. He was of no use, but he was calm. After all, a furious gale is nothing compared to a German mortar attack."

"I like him," I said, "Duncan Winterfinch. I'm not quite sure why."

"When my grandfather died, my mother had wanted to sell the place. But she couldn't. Because he left it to me."

"*You own all of this?*"

"And the surrounding land. Plus Haaf Gruney and a few reefs in the vicinity."

"But why do you live in the stone cottage?"

"I get lost in here. I don't have the . . . *stature* to fill this house. When I was little, I went to public school. I was happy there, in fact. But the rest of my path had long since been set out for me. Edinburgh School of Economics, then heading up a division of Winterfinch Ltd."

"You'll never lack for anything," I said.

"But that's exactly it, I am lacking something substantial."

"And what would that be?"

"I'm like Grandfather. I have no head for business. I have zero interest in import duties and profitability. I sit in the lecture theatre listening to other people ask questions while I hide in my seat. I failed my last round of exams, and it went even worse in the spring."

In the dim light I could at last see what really lay behind Gwendolyn Winterfinch's facade. Pure, unadulterated despair.

"I get money from parents who are disappointed in me. Enough for expensive clothes and music, but not enough to maintain this monster of a house. If I were beautiful, I would have found a husband with a fat bank account. But I have small breasts and I'm terrible in the kitchen. The only thing I know about is antiquities and sailing in heavy weather. The few affairs I've had have been with older married men. Dear Lord, Edward, don't you see? I'm a trinket – and I'm not even attractive."

13

I SAT IN THE TWEED JACKET UNDER A SYCAMORE, DRINK-
ing black tea with honey and reading the *Shetland Times*. On the
table were the remains of breakfast: black pudding, fried eggs and
grilled tomatoes. I had cooked it on the gas stove in the stone cottage and carried it on a tarnished silver tray with a lid through the
chilly air and into the arboretum, where we sat under the trees.

Calm we were not. It was like when *Geira* went to sea. The
undercurrent from 1971 had no end. Never again would I wake
up to Hanne's face. A nice, easy girl, traded in for an affair which
could not end well.

Money – I earned it and I spent it. My share of the income
from the potatoes or the sheep we sent to the slaughterhouse. The
black soil of Hirifjell would refill the coffers next harvest.

But this? I thought, and looked up at Quercus Hall. Great
sums of money had a power all of their own. Before, I had looked
at the house and thought that it belonged to others, that it *was*
someone other. Now I considered what I would have done if it
were mine.

From what the gunsmith told us, the materials for the stocks
– if they still existed – were worth far more now than during the
lean post-war years. I calculated in my head what they could be
sold for, more than enough to buy any large farm in Saksum.

But Mamma had not felt any such inclination, at least not
according to what she wrote to Einar. Despite a beggarly childhood.

There were dregs in the tea, a necessary evil according to
Gwen, who despised teabags. I poured them out over the clumps

of grass, which glistened. A butterfly whirled and landed close by, attracted by the sweet smell.

Who did the walnut belong to in fact? My family in France, who for generations had looked after the woods, or Duncan Winterfinch who both fought for the trees and purchased the right to fell them? That was a question which over the course of the night had become more and more difficult to answer. Because Gwen, having shown me the west wing of Quercus Hall, had clasped her hands together and said:

"We can't just go in without a plan. To be fair I'm a terrible student, but four years at the Edinburgh School of Economics has certainly enabled me to find my way around a business archive. But before we start rummaging, I need you to answer one thing."

"What kind of deal are we going to make, right?"

"Exactly. There's no getting away from the fact that Einar cheated my grandfather."

"We don't know that," I said. "We still don't know why he blamed your grandfather."

She went to a window. With her back to me she said: "Edward, there's only one way to do this. We keep nothing secret from each other. If we find the walnut, we share it equally. And at the same time you might find the truth about what happened in 1971."

Once again we went through the corridors, and she unlocked a steel door and led me down a staircase. The smell of the cellar was heavy and raw. Soon we were in a room as large as the public library in Saksum, brimming with files and discoloured ring binders.

"The business archives of Winterfinch Ltd up to 1947," she said. "Plus all of Grandfather's private archives. Unfortunately a little disorganised at the moment."

A broken ladder lay on the floor. In the corner a bookcase had toppled over and scattered a pile of papers. Only half of the light bulbs were working, but still it was easy to see how the number of files was like a barometer for the rise and fall of Winterfinch Ltd. An entire bookcase was needed to cover the period from 1899 to 1906. Trade dwindled from the twenties, improved again and then dropped in 1940. For the period after the war, when Gwen's mother took over and moved the head office back to Edinburgh, there were only a few files.

"Help me with this, would you?" she said, pulling out a crooked stepladder. "Hold this steady while I go up." She pointed at the highest shelves. "His old private archive."

"Here's something," she exclaimed an hour later. "A medical certificate from the war."

The typewritten sheet was thin and delicate, the full stops had punctured the paper.

> The unit was exposed to a gas attack in position 324 Thiepval/Authuille. Possibly cyanide or arsenic, but widely divergent symptoms (confusion, insubordination). Nearly all soldiers and officers abandoned their positions or passed out. Forty-seven soldiers killed or wounded by machine-gun fire. Thirty-two drowned in the bomb craters. Captain Winterfinch was found at Speyside Avenue near the Ancre, carrying his severed arm.

"Speyside Avenue?" I said.

Gwen knelt on the floor and skimmed the document, then a second one, then flipped back to the first and read it more closely.

"The passages and trenches at the front had street names," she said distantly. "To control the traffic. Thousands of soldiers had to find their way." She handed me another certificate.

> Left arm severed between shoulder and elbow. Carried arm himself and refused to abandon it. The remaining section amputated by a field doctor.

We kept looking, and before long she had found a worn and tattered military map from the autumn of 1916. It unfolded in her hands, and I shuddered to remember how *Det Hendte 1971* had opened for me once.

The Ancre. The village of Authuille. A shaded area above the riverbank. The Daireaux woods. Every single ridge, every shallow in the river, every path and field were clearly depicted. The front line, enemy positions and dressing stations. The line called Speyside Avenue. Symbols and arrows etched in pencil, presumably by Captain Winterfinch before his arm was torn off.

Position 324. Where they had tried to dig in with the machine guns. Where sixteen walnut trees had grown.

The war began to recede from the map, and what remained was the terrain. As it had been before the artillery positions, when the walnut trees began to grow in the 1500s, like it was in 1971, when Mamma and Pappa and I had been there.

Of course I had always known my parents were dead, but where it had happened was only a dot in the atlas. Here I could actually see the place, as clear as an orienteering map. At a small section below the edge of the woods the river curved and spilled out into three large ponds.

They must have drowned in one of them.

An unfamiliar realisation spread inside me. A combination of conviction and anticipation, like when I had been driving around with Bestefar's photograph of Haaf Gruney and watched the terrain gradually fall into place, as though two identical sketches had come into alignment.

A memory was in bloom.

And slowly it happened, the map aligned with my memory. A path we had walked along. The smell of the woods. Strange birdsong. Two warm hands holding mine, one larger than the other, Pappa's. Branches we had to push aside. A sweep of green, a marshy tract. Then nothing.

Had I run away from them then? Or was I imagining this?

Gwen had not noticed that I had retreated into my thoughts. She chattered to herself as she searched through the company archives, which in themselves were proof that Winterfinch Ltd had remained faithful to its suppliers: up to 1929, the archivists had used only pale-yellow binders from Stonehill's, and from then until 1967 they had switched to Eastlight binders with grey marbled spines.

But for the year 1943 a yellow file stuck out amongst the grey. A Stonehill binder with 1921 on the spine, misplaced by twenty-two years.

In it she found a contract, written in French, on the old letterhead for Winterfinch Ltd. *Suppliers of fine and exotic materials worldwide. Edinburgh – London – Rangoon – Georgetown – Takoradi.* The agreement was dated 1921 and concerned the right to cut down "all trees" in the Daireaux woods in Authuille. Next to Duncan Winterfinch's rigid signature was a name in blue ink, the letters compact – *Edouard Daireaux.*

He had been a true farmer, my great-grandfather. For his intention behind the agreement with Winterfinch had not been to make money, but to make the woods viable once more.

For the right to fell the sixteen walnut trees, Winterfinch was to pay for private sappers to continue where the authorities had left off. All explosives and "recognisable body parts" should be removed from the forest floor, and new trees planted.

But it was not long before problems arose. And for the two of us standing in Quercus Hall seventy years later, the same disagreement would bubble to the surface. We began to guard our words, and at the same time looked for slips in whatever the other had to say.

"They were due to receive a generous sum when the trees were felled," she said. "Presumably enough for your family to build new houses on the farm."

"That's of little help," I said, "when it's clear that the contract was never fulfilled."

Among the papers was a report from the supervisor of the sappers which stated that three men had died in the course of a few days. The ground was boggy. The equipment was no good. The problems multiplied. When the gunmaker had described how the patterns and colour of the wood were still developing, that was only half the explanation for the delay. Winterfinch never managed to get a work party to clear the woods.

In fact the work to remove shells went on for years, and even on land where it was possible to use a tractor and plough, hundreds of sappers died. It was not until the thirties that Winterfinch managed to find people who were willing to take the risk, but each team was clearly more ragtag and alcoholic than the next. In the end everyone walked away from the contract. Thickets and bushes had now grown up between the shells, making the undertaking even more hopeless. Winterfinch wrote letters to the Renault factory in the hope that they could speed the development of the new "wonder machine", an armoured tractor with an iron thresher that set off the detonators without anyone getting injured.

But by then the war commission had begun to grumble. They protested at the use of rough machinery on a forest floor on and in which there lay the bodies of thousands of British soldiers. Either they should clear the woods by hand, or they would have to leave the mass grave as it was, behind barbed wire.

We continued our search, went swiftly through the private archive and more slowly through the business files. Here and there we found evidence of Winterfinch's plans for the sale of the wood: An option contract with Purdey, supplier of hunting weapons to the royal family, for the purchase of thirty blanks for an astronomical sum.

Until now Gwen had seemed a little ashamed of the treatment my family in France had received; all she seemed to be uncovering were cold calculations about profitability. But then she found a piece of paper which turned everything on its head. It was a letter to the director of Scottish Widows, a bank originally established as a fund for the widows of soldiers who had fallen in the Napoleonic Wars.

"I assume Scottish Widows still exists," I said.

"Of course it does. Their symbol is the same, a widow with a veil, but she's become a little less sad over the years. Much less sad, in fact."

Then she was silent. I leaned over her shoulder to read.

"Just as I thought," she said. "He had never meant to keep the profits. Look, he'd established a fund for the surviving family of the soldiers. He wasn't going to keep a single pound."

It was now six o'clock in the morning, but we were determined to keep searching to the bitter end, and when we found the next document of any significance – an instruction from Winterfinch to the clearing team – we felt wretched for each other.

The years had passed. Winterfinch had been paying the Daireaux family a tiny sum each year to maintain the rights to the walnut trees. It was not until 1938, with new developments in protective gear and metal detectors, that he managed to initiate effective clearing work.

Winterfinch had stressed in his instructions that only the area around the walnut trees was to be cleared, so that the old paths into the woods would remain perilous until the trees were felled and transported out. The safe areas and access to them should be marked on a map referred to several times in the letter, but which we had not found in the file. The letter closed with Winterfinch's plan to protect the walnut from illegal felling: Gas shells would be left as a perimeter around the trees and along the shore of the river; in effect it should be mined so that no-one could gain access.

Gwen stood with the document as though it were a death sentence. She was silent for a long time before placing it back in the folder.

We left the cold of the basement, passed through the long corridors and reached the warmer entrance, where we were greeted by morning as she slammed shut the door to Quercus Hall.

"Our agreement is terminated," she said, marching off towards the stone cottage. "I don't care what might have happened during

263

the war, or any time since. Obsessions can go from generation to generation. Grandfather spent his entire life casting bitter looks at Haaf Gruney, but this story has nothing to do with *us*, Edward. Come, let's sleep. Or at least go to bed."

We lay naked as the morning light streamed through the curtains, but I could not sleep. Because my deception was already a fact. As she read the documents, I had taken out the Leica and secretly photographed Duncan Winterfinch's old military map.

14

DAYS OF WARMTH. WE WENT INSIDE HER STONE COT-
tage, thick stone walls separating us from the weather. Oak which
smelled faintly of honey, a gentle fire as we listened to music.
Gwendolyn liked to play music loud, really loud, she liked rebel-
lious bands – The Clash, The Alarm and The Pogues – and she
was a completist, bought even obscure maxi singles and bootlegs.
I collected the sheep carcass from Haaf Gruney and we seasoned
it with thyme and rock salt, locked ourselves in and shared a defi-
ant happiness at our self-sufficiency. Got drunk on White Horse
and woke up naked on the sofa.

From the time we awoke until the time we fell asleep, she
seemed genuine. She looked me in the eye, she no longer needed
to pause before answering. That was what I must have fallen
in love with, the fact that she did not conceal anything and yet
remained a mystery. I began to like the way she was different
from Hanne, and I knew that it was not fair. Gwen would roll
quickly out of bed *after being served*, as she put it, slip under the
shower and then get dressed. Hanne would worm herself under
the sheets and talk intensely, tainting beautiful moments with
endless chatter or awkward conversations.

"Listen, dear Edward," Gwen said, taking my hands in hers. "I
like you. *Really like* you. Even when you're dressed like a tramp.
There are still a few days left of summer. We have a car. There's
a record shop in Lerwick, *and* an Indian restaurant. What more
could a couple wish for?"

I liked the fact she no longer concealed her ancestry. Her

contempt for people who bought clothes on sale. Her irritation that the marina would not send anyone to Unst to carry out a discreet repair job on *Zetland*, and instead dropped hints of special offers on new fibreglass boats.

"Horrid people!" she said, and hung up. "They went on about the *price*, not the quality. It's barbaric. And it was white, a white, synthetic boat! Ghastly! Like an announcement to the entire world that good old *Zetland* has been replaced."

"People don't care," I said.

"It's no way to spend old money," she said. "Grandfather bought a new Bentley every other year. Always dark blue and with the same registration number. *Why follow trends when others follow you?*"

She picked up the telephone again and delivered a couple of unambiguous comments to the boss. A couple of boatbuilders came that very same day. When they were finished, we set out on the boat at full throttle, spent hours cruising around the archipelago. *Zetland* appeared to be her most prized possession, even though it was Italian. Because now that she did not need to pretend, she began to show her full pedigree. Her enthusiasm for objects that had improved with age, for pre-decimal currency, for equipment designed for lengthy safaris, for shrewd plans to outwit Hitler, for actions where Shackleton might have presented an example to follow, or which might excuse Scott's delay in reaching the South Pole.

I did not notice until later that I had changed too. One day she stood in the doorway while I got dressed, then went to Quercus Hall and returned with a suitcase. It was packed with neatly folded clothes.

"No, Gwen," I said. "Your grandfather's clothes. They—"

"—are not his. They belonged to my brother. He got fat and forgot about them."

She turned down the volume on the stereo, and Big Country's *The Crossing* disappeared. "They've been hanging here since the last summer holiday the family had together," she said, holding a shirt up to the light. "Turnbull & Asser. Same shirtmaker as the one you borrowed for Dickson's."

"Is this Egyptian cotton too?" I said and stroked the small-checked fabric. Soft, and yet firm and tightly woven.

She shook her head. "Sea Island, 140 threads per inch. Didn't find an exact match in trousers. But of course you can borrow these." She dug out a pair of dark-brown trousers. "Cavalry twill, probably an impulse buy. Goes well with a Herringbone jacket."

"The tailor has certainly done his bit," I said.

"You understand nothing, darling. You can thank the British class divide for *everything* you now admire. Can you name one noteworthy object that's handmade in East Germany?"

"Not offhand, no."

"Without a social class with good taste and plenty of money, there would be no Arbus divan. No Purdey shotgun. No Bentley to turn your head. Not even Indian food would be as it is. Everything exists because someone was rich enough and discriminating enough to reward a gunsmith or a saddler or a chef with an outrageous amount of work."

I put on the clothes and sat down in front of the fireplace with my feet up on an ottoman. I stared into the flames. Work clothes had always been my retreat – outside all day, dig in, keep going, wear yourself out.

It pressed in on me, the certainty that this was stolen time. Every sheep on Unst reminded me of the sheep back home. Leaving the fields, the tonnes of precious seed potatoes, it was the most frivolous and potentially negligent thing I had ever done.

But now, out of the blue, I did not feel like working. I felt like drinking tea, buying records, sitting here in the middle of the day without feeling guilty. Was that how Bestefar had felt in his annual week away, strolling around in a suit tailored by Andreas Schiffer, a concert ticket in his inside pocket?

We went across to Muckle Flugga, to the lighthouse in the north where the cliffs were battered by a white frothing sea. We strolled up and down the streets of Lerwick, got drunk and checked in at a guest house. Gwen taught me to handle *Zetland*, I was skipper on fast trips to Fetlar, where we walked through meadows and took in the scent of andromedas, watched how they changed colour when the wind changed. On to Out Skerries, where we sat with binoculars and studied the glistening otters. Soon *Zetland* became an extension of my own movements, the long row of chrome-bezelled instruments faithfully passing messages between us. Every metre I walked in that tweed jacket took me closer to becoming someone else, but who that someone else was, what type of metal I had in the casting mould, of that I could not be sure.

For the reminders were everywhere, the towering presence of Quercus Hall, the thought of the dress in the coffin. Still I looked out towards Haaf Gruney, at the ghosts waiting for an answer. It was as though Mamma's voice was shouting: *You must not let it go, you must keep searching to the end*.

What would it be like, my leavetaking of Gwen when I said that I had to head home to look after the farm? What would she be left with when the one sweet thing in her life was expended?

But I dragged out the time. I began to observe her gestures, to see if she would reveal something more. It began to get out of hand. Every time I let my eyes linger on Quercus Hall, she would give me a sidelong glance, like a sleeping animal who all of a sudden raises an eyelid. We began to sniff out nuances in what the other said, and if we mentioned the word "walnut" we would instantly become Einar Hirifjell and Duncan Winterfinch.

She was the one who cracked. "I want to go out to Haaf Gruney," she said one morning. "I'm tired of upholstery. I want cold and wet and stone. Here it's too . . . *much*. I don't like the reminders, and I don't like seeing your gaze wander."

I woke up to her movements, and when our eyes found each other, I knew that she had been watching me. Gwen lay on her side, her head resting on her hand, and held a corner of the sheet up to her throat. Here on Haaf Gruney there was no bathroom, no plush towels. Just a tin bowl, a water pitcher and a paraffin lamp.

She did not have the spectre of a blue dress. She had escaped her ghosts, but had not grasped how Haaf Gruney brought mine to life.

"I can't understand why your grandfather didn't fell all the trees while the sappers were still there," I said.

She looked at me stupidly. "Is that what you've been thinking about? They didn't think that the pattern would have finished developing until 1943. How could Grandfather have known the war would come? Anyway, it's a job for specialists. Felling trees is easy, but if they're going to be made into weapons stocks, they have to be sawn up immediately, with rift cuts."

"With what?"

"I did pick up *something* from seven generations of timber traders," she said in irritation. "It requires a specialist, someone who can study the direction of the grain and make the cut exactly where the pattern comes into its own. I imagine Einar learned how to do it at Ruhlmann's. He was just the man for the job. Now let's leave it."

We got dressed and went out on *Zetland*. The weather was calm. There were no storm petrels to be seen. Then I switched on the radio and listened to the weather reports. I could feel the seasons in my body, the instincts of a farmer. If there was bad weather back home, the crop would be destroyed.

"Listen, Edward," she said. "We have to replace the broken window and the kitchen could do with a good clean. I may be clumsy, but I should be able to wash the floor. Why don't you take *Zetland* over to Yell, to the hardware store?"

"You want me to take *Zetland*?"

"Don't worry, she's insured, and the sea will be calm for at least a few hours. I can make the place nice for when you get back. We both need a bit of space, don't you think?"

Why not? I thought. I had covered over the coffin containing the dress and the letters; she would never be able to dig through two metres of peat.

It was so good to be alone on a powerful boat. The tingling in my stomach when I planed it, the propeller's hold in the sea, the vibrations of the motor running through the wooden hull, the wind whipping my face with drops of salt. The sun sharp over the cliffs of Fetlar and the low hillsides on Yell. The faster I went, the better it felt. But for the fact that I could no longer ignore the passenger sitting behind me, silent and trusting, dressed as best she could, her gaze uncertain.

Hanne.

I bought glass and putty and came back to repair the window. Gwen had tidied the pantry and bedroom, cleaned the kitchen and living room. She hadn't done the best job, truth be told, but she had picked some flowers and put them in a cracked coffee cup.

That evening we cast our nets. Eight cod the following morning. The sea was dead calm, there was no movement apart from some geese which had flown over to the island from Fetlar, low over the houses, and landed on the north end of Haaf Gruney, where they plucked at the grass.

I gutted and scaled the fish, went into the kitchen and did not find her there. I called out for her. No reply.

I went out to the steps and saw her clothes on the smooth stones. Her hair floated around her shoulders when she popped up from the water, like an otter.

I took off my clothes and swam out, felt a gentle current take hold. Stretching out my arms, I floated on my back, and she did the same, like a compass with two needles.

"Edward," she said the next day. "I have to go to Edinburgh to sort out some things. And I'd like the marina to lift *Zetland* and take a proper look underneath. I could take her to Lerwick, and then you could pick me up from the ferry in two days' time."

"We could take the car all the way to Edinburgh, like last time," I said.

"That would be somewhat . . . difficult. I have to go to a board meeting, my mother will be there too. She's meeting me in Aberdeen."

That was odd. Gwen had often told me how her mother was both unwilling and too busy to collect her grown-up daughter in the middle of the day. But it seemed she didn't want to trail a Norwegian elkhound about amongst prized English setters. I ought to have cut the tie then and told her I needed to go back to Norway. But then she picked up the keys to the stone cottage.

"You can stay at the cottage," she said. "Do what you like. Take the keys. I have nothing to hide."

"And here, take the keys for Haaf Gruney," I said and gave her the extra set I had got from Agnes Brown.

The red telephone box was fresh and shiny in the rain. Behind the glass panes a dull, yellow light shone.

I sat in the car feeling terrible. I stared at the dry-stone walls, counted the stones one by one.

It was like the time I had to put down Flimre, our old cat. I loaded the gun, stroked him. How innocent he was, present in the life he wanted to live despite no longer being able to eat. I drew out the time, slow seconds remembering our years together, his eyes suddenly realising there was an executioner in me.

Hesitate and let time pass, give up and put the gun away.

Keep on living, keep on stroking the soft fur.

But then I picked up the gun again quickly, before my own disbelief in what I was doing got the better of me. I pointed the barrel of the .22 at his head and fired. Watched as his body was thrown to the ground, and held him in my hands as the cat's life shuddered to an end.

After half an hour of denial, I climbed out of the car. In a brief break between waves of regret I grabbed the receiver and dialled the number for Hirifjell.

"There you are, at last," she said. She talked about everything she had done on the farm, how much she liked it, the flowers in bloom, the food she had eaten that day. "You said you were not going to be the same," she said, talking non-stop so that I couldn't get a word in. "I've changed too. I dyed my hair."

Then it petered out.

We searched within ourselves for the right thing to say.

"Hanne," I said. "Dear Hanne. I asked if you could check in on the farm. The fact that you're living there, that's totally fine. But—"

"But what?"

"I don't think we should be there together when I come home."

For a long time she was quiet.

Then I heard a girl crying at Hirifjell. A girl holding a heavy receiver in front of a chest of drawers on which was the photograph of Mamma and Pappa. Three of the four people I had really been happy with, and still I did not know if Gwen could be the fifth.

The next day I stood among the gulls at Holmsgarth and watched the Aberdeen ferry glide slowly towards me. Ten minutes later she walked out of the terminal building and down the steps in a dove-grey outfit with newly styled hair. I only wished it was Hanne there, Hanne with her honest smile.

But here was Gwen, and the rope which tied me to Hanne split faster than I had believed possible, a thin thread which snapped when Gwen threw herself at me and I took her full weight in my arms. She kissed me on the mouth and weakly I allowed myself to hope that we were a couple, that my betrayal of Hanne would be justified by true devotion to Gwendolyn Winterfinch.

Perhaps I should have been more surprised that Gwen did not smell of salt water but of coal, like the chimney smoke in Lerwick that seeps into your clothes.

My own betrayal must have overshadowed any suspicion. Because while she was away I had let myself in to Quercus Hall, where I spent hours searching through the private archives.

*

I dropped to one knee in the drizzling rain with a slender Dickson Round Action. It felt alive, responsive. The steel was cold, the wood just as cold, but somehow felt warmer. I opened it, dropped two heavy orange cartridges into the chambers. The sound of a reversed sigh as they slipped in, followed by a metallic click as the brass head struck the steel of the barrel. Two Eley Grand Prix no. 2 from the box in Einar's wardrobe.

Two shots which would ring louder than any other shot. Two shots to see how Gwen would react to a reunion with the walnut which divided us.

I heard the squawking of the geese. They had changed direction, a sign that the seasons had changed too. A small flock slipped over the grassy ridge of Fetlar, crossed the inlet and approached. Some climbed high on the wind, but two kept low and came so close that I could pick out nuances in their plumage. When I could hear the flapping of their wings, I raised the shotgun.

Never had I felt anything like it. The gun I had inherited from Pappa was like a piece of driftwood in comparison. The Dickson danced, slipped into position as though it was a part of my body. The slender barrels followed the bird, and I barely needed to think for the gun to take care of the rest automatically. The recoil was little more than a report to the shoulder that the gun had been fired, and I saw the feathers fly as the shot struck home. The goose flapped its wings once more and swooped diagonally downwards, in more or less the same direction as its original flight, before crashing to the ground already dead. My thumb ran down the beautifully cut steel, the spent shell was ejected in an arc, with a trail of gun smoke.

Gwen walked bare-legged in front of me.

"Brilliant shot," she said.

"The report would have carried a long way," I said. "When is the hunting season here?"

"Who would anyone complain to? The owner of the land hereby gives you permission."

I placed the shotgun on a flat stone and let her comment go.

We were just two people in the mouth of an inlet which had yielded up food; I wasn't the one with dirt under my nails and she wasn't born with a silver spoon in her mouth, we were just *us*, hungry and cold, with a goose which steamed as I cut it open.

But the unease forced its way in. The night in Quercus Hall had been fruitless. I had not managed to wrest more secrets from the archive room; I had simply studied the war map and felt the need to travel to France grow, to leave everything and travel south.

Gwen looked away from the entrails and my bloodied hands. Her eyes fixed on the shotgun. In the morning sunlight the wood had taken on yet another appearance, like a painting which becomes more difficult to understand the longer you look at it, until you are forced to accept that it is unreadable.

"You should put some clothes on," I said.

Gwen picked at her food, moved her fork in strange, elliptical lines across the plate. The freshly fried goose tasted of almost nothing. A smell of weapon oil had settled faintly in the room. I had cleaned the shotgun before we ate, and Gwen held the cellar hatch open while I slipped it into a dry hiding place.

"I should have hung it for a few days, this one," I said, chewing.

"Or a few weeks," she said mutely and pulled a fibre from her teeth.

I asked myself when we were going to argue about what we *really should* be arguing about.

"By the way," she said. "I was up by the crag. Saw something strange over on Unst. Someone has painted a white X on the boathouse."

15

TOO LATE. *GEIRA* WAS ALREADY HALFWAY ACROSS THE inlet. My fingertips were still white from the fresh paint on the boathouse. A brush had been left on a stone.

Gwen had taken me across in *Zetland*, still in a grumpy mood. I had forced myself to finish eating before telling her that I wanted to photograph the lighthouse on Muckle Flugga in good light, a task so tiresome that I was certain she would not want to join me. We parted with a sullen agreement to meet in the stone cottage that evening.

I stared at the stern of the ferry as the Commodore idled. But Unst was like Saksum. There were only two or three places where you needed to look.

She got up from the bench outside the cemetery in Norwick. Her white hair shone, she had it up and was dressed in a black loden cape with red lining. A thin silver armband on her wrist. She seemed more frail than the last time I had seen her.

"I forgot about the key in the coffee tin," she said. "Did you find it?"

"The key?" I stood with my hand on the car door.

"Einar rented a storage shed in Lerwick. The key was in a tin in the larder."

"What was in the shed?" I said, walking over to her. "Do you have any idea?"

"Materials, I would think."

"Might it be a load of wood?" I said.

"Could be. I assumed it was planks and boards he brought off the Bergen ferry."

I should have sat with her on the bench, out of politeness, but I was impatient. I experienced a shabbier version of what Einar must have felt; if a missing key made me so uneasy, how the need to find a lost lover, and a lost daughter, must have eaten away at him over the years.

"You're restless," she said. "I can see it. You want to go across and look for the key, don't you? Do you know why I painted the cross?"

"Wasn't it because of the key?"

"A short time ago," Agnes Brown said, "I was told something by the doctor. So if there's room on the boat, I would like to see Haaf Gruney once more."

We opened up the boathouse and pulled out *Patna*. A white-haired lady dressed for the weather, and I like a phantom of a forty-year-old love. I wondered if Gwen would see us and follow.

"You are too polite," Agnes said suddenly.

I continued rowing and shook my head, puzzled.

"You didn't ask about the doctor's message."

She pulled the cape around her. Her eyes wandered to my hair.

I began to tell her about the walnut Einar had hidden from Winterfinch. She nodded patiently at first, but then she began to look uncomfortable, as though she were waiting for a particular revelation. There should have been no harm in Einar telling her everything, but he had not. Yet another reminder that he had not appreciated her. I regretted having spoken.

A wind raced across the sea and rippled the water. Gulls flew high above us. Haaf Gruney drew nearer with each pull of the oars, and all the time I kept my eye on the shining white cross on the boathouse.

She could tell what I was thinking.

"What would you do," Agnes said, wriggling to get more

comfortable, "if you found the wood and managed to sell it for all that it's worth?"

"Maybe buy back the farm in France. If that's what Mamma would have wanted. But I suppose I'll never find out."

We came ashore and walked through the rocks and yellow tufts of grass. An older lady stepping cautiously over slippery stone. All she saw was an emptiness. I looked nervously over towards Unst. At any moment *Zetland* could plough a white furrow through the sea.

"I would like to find a way to thank you properly," I said.

She drew the cape around her and looked down at the stone buildings. "Back home in Ørsta I don't have much family. I don't know many people here anymore either. So I have one wish – that you sing *Kjærlighet fra Gud* by my coffin."

"That one up there," she said, pointing at a shelf. Behind a cracked iron pot was a red and white tin – Norwegian Ali coffee.

"I bought it when I was in Førde one time," she said. "Thought it would cheer him up. He pretended to be pleased, but he had no feelings left for Norway."

I reached for the coffee tin. There was no rattling when I shook it. "There's no key in here," I said, removing the lid.

"That's strange." She looked down at the tarnished metal. "I checked it was there when I locked up the last time."

I stood on a stool, picked up the iron pot, moved aside unlabelled green bottles, opened another tin. Found no key.

"The address of the storeroom was on it," Agnes said. "Imprinted in the metal."

Her gaze moved around the room. She peeked into the kitchen and saw the goose carcass and two plates with food remnants. They were probably the same plates Einar and she had eaten from.

But she did not ask who my guest was. I was hardly able to admit the name to myself: Gwendolyn Winterfinch.

What had she said a few days ago? *The kitchen could probably use a good clean.*

277

I got down from the stool and thought about the day she had returned, supposedly from Edinburgh. But she had had the smell of Lerwick coal in her clothes.

How sly was she, really? Tell me this is a coincidence, I prayed. Tell me that she's sitting in the stone cottage right now, that when I knock everything will be like it was before. That she found the key and left it somewhere when she was tidying up. Let Gwen simply be a young girl holding her grandfather's hand, who had to let it go for good when he died.

"Agnes, where's the storage lock-up?"

"In Gremista Brae, near Holmsgarth."

"Gremista – what's that?"

"A small industrial area."

It was a warehouse for everything rough and rusty associated with fishing boats. Five long, windowless, corrugated-iron sheds. *Ring here for service*, was written in felt-tip on a scrap of paper. An arrow pointed downwards at a dirty buzzer.

The guard was younger than me, a red-haired youth in a grey Beaver-nylon boiler suit. A long-handled Maglite hung from his belt along with an enormous ring of keys.

"I'm here to pick something up," I said, "something Einar Hirifjell stored here."

"Do you have the key?"

"Unfortunately I don't, but here's my passport. If we could work something out. I'm his next of kin."

The boy led me towards a shed, reached through a sashed window and pulled out a curled-up record book. Shook his head, went inside and found another. He reminded me of the type who directed ferry traffic.

"What are you here to pick up?" he said.

"I don't know."

"You don't know what you are picking up?"

"Presumably a shipment of timber. Some materials he had stored here."

He flipped through the record book.

The last couple of hours had not been good for the nerves. I had asked if Agnes wanted a lift back to Lerwick, but she said she would take the bus, as usual. On the ferry landing I stood thinking about what Gwen and I had agreed. To share equally.

Then I had gone to the stone cottage, thinking through how I would ask about the key.

But Gwen was not there. Her footprints led through the grass up to the cottage, but the blades had risen again. I had run down to the boathouse. *Zetland* was gone. A faint smell of exhaust indicated that she had left only recently, but in the inlet there was no boat to be seen.

"Nobody by that name has anything here," the security guard said.

"You're sure that—"

"Sorry. Nobody with that name."

He handed me back my passport. I just stood there while my thoughts whirled.

"What about Oscar Ribaut?" I said.

I could see his scepticism grow. He leafed through the yellowed pages again, stared at one for a while. He stepped away from me to exchange a few crackling words on his walkie-talkie.

"Do you have a sister?" he said, turning back.

"Me? No. Why do you ask?"

"I spoke to the woman who's normally on duty. She said Ribaut's lock-up hadn't been opened for years."

"Wait a minute," I said, "so Oscar Ribaut actually had something stored here?"

"Still has. He paid the rent for ten years up front. And there's a clause here too; if you don't have the key, you can answer three questions to be let in. But there might be no point – a few days ago someone came to look at the items."

"And who was that?"

"A young woman, apparently. Nicely dressed. Edinburgh dialect. She had the key."

So she had thrown herself around my neck having betrayed me. The strength within her that could master rough seas, that made her open up *Zetland*'s throttle fully, her entire inherited determination must have bubbled to the surface when she found the key.

But had we not been together for days afterwards? I wanted to feel anger, and contempt, but all I could picture was her innocent face on the veranda of the arboretum.

"So she took everything?" I said. "The girl."

"Don't know. I wasn't here that day."

"The clause," I said. "What was it? The three questions."

He glanced at the yellowed page.

"Eh. The first is whether you're wearing clean underwear."

"Is this a joke?"

"It says I have to ask the question!"

"Well, they're from yesterday. Was that the right answer?"

"I don't know. It doesn't say."

"And then, what's the next question?"

"Whether your name is Edvard Daireaux Hirifjell. But I see from your passport that you have two of the names at least."

"Yes, yes. And the third?"

He stared at the sheet for a long time. "I can't pronounce it. It's in German or something. *Waa hitter preston soh confirm die hem i Sachum?*"

In my head I tried to spell out the words on a board.

"What's the name of the priest who confirmed you in Saksum," I translated slowly.

"Yes," he said. "That sounds about right."

"Thallaug," I said. "Magnus Thallaug."

He pulled up his key ring. The metal rattled for an eternity before he picked out a small key and led me towards a shed.

The warehouse was cold and poorly lit. We walked past old forklifts, a torn cardboard box of wellies, past outboard motors, ropes and fishing boxes before we came to some wire-mesh cages.

A ceiling light cast a beam on a cage containing something covered in grey tarpaulin, four to five metres long and with an elevation in the middle. Someone had been in recently and left footprints on the dusty cement floor.

"Heh. So she couldn't have taken everything," the guard said. "Well, I'll leave you to it."

His steps receded and I walked into the light. The tarpaulin was covered in a thick blanket of dust, which had slipped off where she had lifted the cover to look underneath. I took hold of a corner and caught sight of a dull hubcap and a punctured tyre.

A car.

So that's why. That's why Gwen had smartened herself up, that's why she'd thrown herself around my neck. She was happy that the walnut was not here, happy to avoid being dragged to the very depths by an inherited obsession. She had probably stayed overnight in a hotel. She had stepped out of the ferry terminal when the Aberdeen ferry docked, and she had actually been *happy*.

So why had she taken off this time?

I rolled the tarpaulin off the roof of the car, sneezing as the dust swirled. The car was worn and dented, and submarine-grey. Possibly from the early sixties. No emblems. The front pocked from the spray of gravel, the driver's seat worn and sunken. The bonnet was long, with an indeterminate design. It would not stand out in a car park.

The keys lay on the seat. The dashboard was crowded with instruments and rear-light switches and resembled the cockpit of an aeroplane. It was only when I saw the BRISTOL emblem on the steering wheel that I knew what make of car it was.

We pushed it into the sun, the guard and myself. It rolled heavily and reluctantly on deflated tyres, creaking with every movement, the wheel bearings sluggish with congealed grease. We straightened up to catch our breath and looked at it standing side by side.

Its wounds were more visible in the daylight; it was scratched up as though straight from the breaker's yard. The musty documents in the glove compartment showed that it was a 406 prototype with a V8 engine, "approved for road use" and sold with a certain degree of damage present. In the space for the purchase price were the words, "Shop-fitting work at 368 Kensington High Street, as agreed with the proprietor, Tony Crook."

"The steering wheel's on the wrong side," said the guard, pulling out a cigarette.

I was about to disagree with him but then remembered where I was. It was left-hand drive, made for driving in Europe, to search for a child born in Ravensbrück.

"So everything's in order then?" I asked.

"For our part, yes. You still have rent in your favour."

"Can I exchange that for the use of your rear courtyard?" I said. "I'd like to get the engine running."

He flicked the cigarette and gave me the thumbs up.

The Bristol was a combination of the best of American and English, a machine in aluminium and cast iron that would go on for ever. Stuck to the front air duct with yellowed tape was a note: *Check brake fluid.*

Was this the only message Einar Hirifjell had left? And who had he left it for? Himself? Me?

I opened the glove compartment and found a receipt for an oil change carried out somewhere in Germany in 1961. There was another from Czechoslovakia from the previous year.

I pictured him arriving at ruined churches with his chisel and plane, repairing pictures and statues of a God he must have believed in less and less, and then more and more. The car must have driven many thousands of kilometres. Driven by a man who did not belong anywhere, who was restless everywhere but at the workbench.

In the boot, wrapped in brown wax paper, were some spare parts and a tool box. My pulse quickened when I spotted the mouldy leather case of a camera. Inside was an Ilford Witness, good as new, but it contained no film. A tattered road atlas for

France, Michelin 1948. I looked up Authuille. No arrows, no ring in pen around any city. Just equally thumbed pages.

On the floor mats I found shavings of white birch bark. A ferry ticket from Bergen to Lerwick with Smyril Line in 1978. My tenth birthday. Perhaps the car's final journey.

Take it easy now, I interrupted myself. Imagine yourself in his place. What would you have done when you parked the car? A letter, he couldn't have left that. Not even here. Why? Because the plan had gone wrong; someone was still searching for the walnut.

Gwen. I tried to invest her with what I *thought* she was, retrieved memories of our best times, days which until now had been mementoes, but which all of a sudden had changed their appearance.

Think, I muttered. What kind of signal does it send, leaving a car equipped with spare parts and a road map of France. Perhaps the inheritance is not in the Shetland Islands after all.

I took the battery from the Commodore, filled the tank from my spare petrol can, and then I offered the engine some starting fluid. Evening had descended on Lerwick by the time the Bristol hummed to life. The air had turned chilly, my hands were filthy and I had not eaten for hours. I got into the driver's seat and tried the gears.

In the glove compartment was a white eight-track cassette: Glenn Gould Plays Bach: The Goldberg Variations. Bestefar had had the same, by another pianist. I pushed it into the player and climbed into the back seat. The fan gave out a nice heat, and the leather seats warmed around my body, emanating a scent that was foreign yet familiar.

With a jolt I sat up. The warning came not from the brain, but from my senses, like an impending earthquake. Like the moment I knew that Bestefar was dead. Like the shock when I realised that Hanne had moved in to Hirifjell.

This car. This smell. The pattern in the roof lining. The wear on the floor mats. The scent of leather and age. Glenn Gould's piano.

My gaze fell on the door of the glove compartment; it did not have a handle, just a small strap of braided leather. Why had I not remembered that detail before? Because I *knew* that strap.

Suddenly I was back there, sitting in the back seat terrified, as terrified as the time I heard the cracks in the flame-birch woods. I was so small in the seat, and the car was going at high speed, and then I remembered something else, I remembered the thin back of a man in the driver's seat. He said something which was meant to calm me, but the words had no effect, because I was looking for something. Something made of finely polished wood.

My toy dog.

I was looking for my wooden toy dog at the same time as a wall of dark trees passed outside. Then my hands remembered something else, the feeling of a thin fabric. The dress in the coffin out on Haaf Gruney. *Had Mamma sat in this car?*

Had I lost the toy dog, or had I lost my parents?

Then the memory faded, but the certainty remained.

I had sat in this car before. During the four days of my disappearance.

16

A FAST DRIVE TO UNST IN AN UNINSURED BRISTOL. THE tyres were cracked, the steering wobbly and blue smoke blew out of the exhaust pipes. But something was happening, there was a connection between me and this car which grew stronger the further I drove. The needles on the instrument panel quivered, the passenger seat vibrated from the imbalance in the wheels and hazy impressions from the past raced by like animals in the dark.

Quercus Hall towered in the night. *Zetland* was nowhere to be seen and Gwen's stone cottage was still empty. No-one in the vicinity, and not even the baying of sheep, just the wind howling across the dry-stone walls.

I rowed out to Haaf Gruney. Opened the cellar hatch and groped for the shotgun bag, but it wasn't there. Gwen had used her keys here too.

The deception had been successful then, on the second attempt. In the outbuilding I dug up the coffin from beneath the peat, then washed my hands and moved aside the lid. I unwrapped the silk paper and stroked the dress. It was familiar to my hands, the memory was true.

I rowed back to Unst and slept in the Bristol. Let the smells seep into me, and with them the hope of more memories. Like a seed testing whether the soil is fertile. I woke to the sight of Haaf Gruney across the water. Low and dark, like a worn gravestone.

For a long time I sat with my eyes closed. Further memories had not surfaced.

The smell of the leather seats *was* familiar, but I could not remember more; it was like a drawing had been coloured in, but it was still the same drawing.

In the morning light I turned the car inside out, but found only dust and coins, all from the fifties or sixties: German pfennig, Czech haleru, French centimes. Until, between the seats, I found a familiar shape – a spool of film.

Laing's Pharmacy in Lerwick opened at nine o'clock. I waited outside, studied the fishing boats by the pier and the passing workers.

Gwendolyn, no doubt she was in Edinburgh, cashing in thousands of guineas for a Dickson Round Action, "Attractive to us, for a discreet transaction". Her prize of the summer was an adventure in which she duped a stupid Norwegian, the fulfilment of a romantic plan she had formed as a teenager.

The bell jingled. I walked inside and greeted the chemist, a fair-haired, strikingly good-looking woman in her fifties. The place was living proof of a time when a chemist was the closest connection between the chemistry of photographic processing and photography itself. She had a small shelf of Kodak and Ilford film, and a couple of Olympus cameras in a glass showcase.

"Is it possible to get a film developed today?" I said, putting the film on the counter.

"I'm afraid it would have to be sent to Aberdeen – we no longer develop ourselves."

"Damn," I said to myself and shut my eyes. When I opened them again, the chemist was holding the film in her hand. "Hm," she said. "An Orwo NP20. Seldom see these."

"It's old," I said.

"I can see that."

"When would it be from?"

She held it between her thumb and index finger. "Late sixties, maybe?"

I looked at the black developing tanks on the shelf behind her. Paterson, the same as I had back home.

"Do you have the right developing powder?" I said.

She looked it up in a book. "Unfortunately not. The Orwo is rather particular. I have Ilford Microphen, but it won't be very good. You could use Rodinal, but a freelancer at the *Shetland Times* bought the last bottles yesterday."

"I can mix it myself," I said. "Do you have hydroquinone?"

Perhaps I had not pronounced it properly. She took off her glasses and gave me a searching look. Seconds later my eyes had supplied the answer to a question she never asked. From a drawer she took a chart, then fetched some empty brown bottles and began to write out labels. "Natrium sulphate, we must have that. I also recommend calcium bromide, to reduce the haze of ageing. Don't you agree?"

"Absolutely," I said.

She drizzled powder into a small bag. "This is poisonous," she said, sticking on an orange warning label. "Are you sure you know what you're doing?"

"At least until the film is developed," I said.

Out on Haaf Gruney I collected rainwater from a pool. I filtered it in the kitchen using a sieve, tasting it to be sure that no salt had blown up from the sea. I loaded peat into the kitchen stove and fired it up, then crawled down into the cellar, making sure that no light could come through the cracks.

I felt the magic of breaking open a roll of film, the certainty that there was something fragile and alive on the light-sensitive silver. Invisible now, and locked in another time. I realised that was perhaps why I went numb whenever I had a film in my hands in the darkness, because it could capture time, and time was something I had once lost.

I wound the film into the spool, placed it in the developing tank, opened the cellar hatch and climbed back up.

This is important, I said to myself. You only have one shot at this.

I heated the water and placed the thermometer I had bought

from the chemist into the cooking pot. Too hot. A little more cold water. There. Twenty degrees.

Quickly I mixed in the developer and filled the tank, knocking it on the worktop to get rid of the air bubbles. I sat down and waited. No way back. Eleven minutes, not ten, not twelve.

I prepared the rinsing bath and dissolved the fixing salt. Kept an eye on the time. Every third minute I turned the tank and gave it a gentle tap.

There. Time to pour it out. The developing solution was darker. A good sign.

Rinsing bath, fixing bath. Another rinse.

Then I took a breath and screwed open the tank.

The film spiralled downwards, drops of water splattered against my hands.

Dangling near the floor were some frames which contained images. They were diffused by a milky veil, but forms were still discernible, and when I held up the film to the window, I saw that the silver had been faithful to a light which fell in France in 1971.

I had no darkroom and no photographic paper, not to mention an enlarger. But I did have a cellar, glass shards from the window-pane smashed in the storm, a torch and an objective lens I could turn upside down. Holding a precarious pile of shards and expensive German optics, I tried to sharpen the images by projecting them onto unplaned wood. The woodgrain was like a watermark beneath the images, which were clear only when I managed to keep my hands steady. They existed only in the moment, a piece of reality from September 1971.

Einar could not have been a skilled photographer, because the first pictures were either overexposed or blurry, or both. The first I could make out was of the Bristol on a ferry quay.

Fourteen pictures from the trip to France. Mamma, Pappa, Einar and me. They must have taken turns using the camera. Us in front of the Mercedes, then in front of the Bristol, perhaps at a meeting point they had agreed.

In the next photograph we were in a lay-by. They must have got someone else to take the picture, because we were all together there, the four of us. A family beneath a parasol with a Cinzano ad. Simple, genuine holiday pictures.

At last, a close-up of Einar and Mamma together. An infinity of torments had lodged in his face since the passport photograph from 1943, but it *was* him, furrowed and scored like a workbench, but the eyes were calm, and he had a hand placed warily on Mamma's back. Mamma smiled thoughtfully, some of her hair hung loose at her cheeks, her gaze directed at the camera.

The next was of me and Einar together. My hand in his, but I was looking at my other hand.

At a toy dog.

So it was real. My memory tried to connect with what I could see, but never grew clear. But still it was something I sensed, a bony hand, stubble against my cheek.

In the next I stood alone in front of a brick wall. Holding my toy dog and smiling. A photograph intended for Einar's barren living room, perhaps. Could I have got it from him?

I shifted my weight and tried to keep still, leaning against the cold stone. It was like sitting inside an empty grave, looking at photographs of the dead. Just fourteen pictures of a simple trip. On a 24-exposure film. Perhaps the trip was not intended to be very long. Or it was interrupted.

When I came to the last picture, I was unable to hold the glass, lens and torch all together any longer. I lost my grip, and the image disappeared.

But I had seen enough.

Mamma and Pappa. I could not make out anything other than the grey tones of their clothes, but the contrast was good: she was wearing a dress with white edging at the collar.

I packed up to leave. I fetched the dress, the chessboard and all the documents. Rolled up the film and placed it in the inside pocket of my anorak.

Then I sat looking at the empty walls.

Gwen had ploughed ahead, oh yes. But it was my fault that Hanne had been left in the ditch. And yet without Gwen, I wouldn't have got anywhere. I wouldn't have met the gunmaker, nor would I have discovered why Winterfinch was so desperate to find the walnut.

These were perhaps growing pains that I could feel, along with the loss of the Dickson Round Action with a stock worth the price of a new Jaguar.

One thing was certain, though. I also felt loss, it itched intensely within me. I had to get away from here, I had to get to France and discover everything that I did not yet know. But first I would have to make a necessary detour, to Hirifjell, to prepare for the potato harvest. And perhaps show my face to Hanne and ask her what could be read in it.

I threw the rest of the goose to the gulls, washed up the two plates and lay down to sleep. The Bergen ferry would leave the next morning, giving me just enough time to pick up the Commodore. Outside, the wind had picked up; I hoped that the last rowing trip in *Patna* would not be too perilous.

In the middle of the night I sat up in bed, woken by a nightmare in which the entire potato crop had failed.

I thought I heard a boat engine near the island, and then the sound disappeared. I lay down again but continued to listen. For as long as I had been alone on Haaf Gruney I had heard nothing but my own footsteps, the wind and the breakers on the rocks.

Now I could hear footsteps. Someone was opening the door in the hall.

I lit the paraffin lamp and pulled on my trousers.

She came in and placed the shotgun case in the middle of the floor. Her hair was wet and flat against her head. Her clothes were crumpled, her fingers black. A patch of oil was spread across her jacket.

I walked past without touching her, out to the front steps. *Zetland* was moored at the pier.

She sat down heavily on the kitchen stool with her back to me.

"Who do you want to be?" she said.

"What do you mean?"

"Do you remember how we were in the car to Lerwick? On the way to Edinburgh? At the restaurant?"

"That was before we told each other who we really were," I said.

"We managed it for a while, didn't we?" she said and unbuttoned her sailing jacket. She was wet to the skin.

There was so much I should have said. How she had been good for me, but had wounded me all the same. But I couldn't find the words. My fondness for her flared again when she said:

"Tell me one thing, Edward. Back when we spent all that time together. Did you like me then? I mean really *like* me? As you would a girlfriend?"

"Yes," I said. "I did."

"And I liked you as I would like someone who could become my boyfriend. And now?", she said, lifting her leg out of the puddle of water that had collected from her wet clothes.

I did not answer. I opened the shotgun case and pulled out the weapon. Someone had rubbed oil on the barrels and waxed the stock.

"I was at Dickson's today," she said. "Caught a plane. They pulled out the chequebook straight away."

"Then why didn't you sell it?"

She took a pack of Craven A from her pocket, but it had got wet, so she threw it casually into the bucket by the stove.

"Haaf Gruney isn't going anywhere," she said. "It will always be here to remind me that I did something wrong. It – no wait, don't say anything, Edward. There's more." She pulled out a key and handed it to me. "I lied. I was never at any board meeting. I was in—"

"Forget it," I said. "I know everything. I've already been there. They told me about you."

She blushed. Shame, biting shame, was something I had never thought would be evident on Gwendolyn Winterfinch's worldly face.

"I was trying to get the upper hand," she said. "But I despised myself afterwards. Then there was that strange cross on the boathouse on Unst. And you said you were going to Muckle Flugga, but instead you rowed back out here with the white-haired old lady. I came over here later, thought that at least this shotgun should be mine."

"What you didn't find that time," I said, holding up the keys for the Bristol, "was the key to the other storage unit. It was with these."

"*What?*"

"Yes," I said. "Einar had another storage unit."

"Were they there? The gunstock blanks?"

"All of them. Nearly three hundred. Of the finest quality."

She stood and zipped up her jacket. "So it's over then."

"I guess it is. Unless you want to call your lawyer."

She walked towards the windowpane we had fixed. Pressed her finger into the putty, which was still soft.

"Sell them," she said. "I don't care. Grandfather is dead. The war is over, the fund with Scottish Widows is closed."

"How would I get the best price?" I said.

She laughed bitterly. "I've planned that out a thousand times over the past few days. You deliver a small shipment to a couple of auction houses, say Bonhams and Sotheby's. Get them to tell a newspaper – preferably *The Sunday Times* – that the lost walnut has turned up. They'll dig through the archives and put their best writers onto it. Oh, that will drive up the prices. Mysteries across two world wars. A missing fortune, a hint of deception which you have to dress up as best you can. Then you contact Dickson, Holland & Holland, Purdey, and not least Boss, and sell directly to them. Westley Richards too. Make it clear to everyone that this is the *entire* shipment. There is no more. Put it about that you're the only descendant of the Daireaux family. Donate a sum to the war memorials at the Somme – just to give yourself a slightly cleaner conscience. I won't interfere."

"You're being honest with me now?"

"Just get rid of it. And I'll wander around Quercus Hall for the rest of my life putting buckets under the roof leaks."

"I'm leaving now," I said. "So I won't see you again."

Her body could find no balance; it was as though she could no longer tell the difference between starboard and port.

"My lectures start next week," she said. "Another autumn in which economics student Gwendolyn Winterfinch will have it hammered into her that she is no good. Mornings when she leaves the lecture hall early to go comfort-shopping for records and clothes. Board meetings in which she doesn't utter a peep. Back to the flat to fiddle with books that don't interest her. Back here to sit alone in Grandfather's office, looking out at Haaf Gruney."

I closed the shotgun case, found my set of keys to the stone cottage and gave them to her. In return she handed me her bunch. One set of Norwegian Mustad keys.

"And now?" she said, walking towards the door.

"Now I'm going to Norway to harvest my crops. Then to France."

She frowned.

"Why would you go to France if you've already found the walnut?"

"Because I still don't know what happened in 1971. And now I have the means to travel."

"Well, have a nice trip," she said.

I watched her walk down to *Zetland* in the half-dark. The jerking in her shoulders was unmistakeable.

I began to cry too. I *had* liked her. I had even liked her when she lied. I liked her lies because they had brought me closer to the truth about Mamma and Pappa.

I followed her down to the boat. I could hear her crying between the sound of the waves.

"The walnut wasn't there," I said in the end. "There was no other key."

Had she continued standing there with her back to me, concealing new calculations, putting together a new plan, then I would have chosen differently.

But instead she turned immediately, and even though I could not see her face clearly, I noticed that her movements were lighter, as though she had cast off a heavy load.

"Then there's still hope," she said, running towards me. "As soon as I get away from here, I'll be freed of all the nagging thoughts. Let me prove it to you."

"There's nothing else you can do," I said.

"Oh yes there is," she said, and she clung to me as she had at the ferry terminal in Lerwick. "I can miss the start of the semester at college. Let me come to Norway with you. Let me come to your farm, Edward. Let me be the silly girl who liked *Forever Young*."

IV

❧

Unexploded Shells

1

ON THE STONE STEPS OF THE LOG HOUSE STOOD HANNE
Solvoll in a white dress. So resplendently pretty that it gave me
a shock, both from the reunion and from seeing how lovely she
looked, with her golden hair and glistening brown skin from days
working in the hot sun.

I had sensed danger the moment we turned off the county
road. The gate was open. The wayside cut with a scythe. The sun
shone on the houses and the lush farm. It was the same sight
which had always greeted me when Bestefar was alive, everything
well kept and fresh, the carrot stalks erect in newly weeded beds,
the kitchen garden red with ripe currants.

This was not how it was supposed to be. The grass should have
been bushy, the vegetables ousted by weeds. When I drove into
the yard and saw the Manta, I squirmed in my seat with regret.
Because Hanne stood in the doorway, no doubt wondering why a
strange Bristol had arrived at Hirifjell. Her hand rose halfway to
wave when she recognised me, but dropped when she saw that I
was not alone in the car.

The trip had been fine, carefree even, and I really believed that
Hanne would be gone. Stupid and naïve, I realised now. It was
as though my sense of reality had been scattered on Shetland.
I considered the trip to Norway no more than a pit stop on the
way to France, and at first I deeply regretted allowing Gwen to
come with me, but this eased as soon as the ferry set sail from
Holmsgarth. All she had brought with her were two old leather
suitcases. My luggage was a chaos of cardboard boxes and plastic

bags. "I suppose you've never heard of 'fitted luggage'," she said. "Suitcases made for the boot. We had them for the Bentley, and I was only allowed to fill these two."

The juddering, creaky Bristol, she fell in love with it straight away, and when she got comfortable on the cracked leather seats, she said that it was like Whitehall on wheels. I enjoyed coming ashore in Norway, seeing the fir trees and proper wooden houses. Driving home with temporary licence plates, drinking fresh water from the tap, seeing the familiar selection of confectionery at the petrol stations, filling up with overpriced petrol. When we passed Laugen in the blinding sun, she said "This place is just marvellous", and she really meant it, right up until we arrived at the tidy farm surrounded by woods.

And there stood the warrior queen in a white dress on the stone steps, like a statue moulded to perfection, who seemed to be saying *You've risked all this, you imbecile.* Hanne scrutinised the Bristol, scrutinised me, scrutinised Gwen. Measured her from the shoes upwards with an expression that said: *Look what the cat's dragged in.* She turned to me, ran a finger across the tweed jacket and said: "Nice jacket. Welcome home." Then she walked to the Manta.

I began to go after her, but stopped myself. I stood in the middle of the yard, between a chubby girl in a Burberry cape and a supple child of nature.

I waited for Gwen to shout, "Who the hell are you?", her Scottish dialect reverberating across the yard. But somewhere in a refined, upper-class upbringing there must be a blueprint for situations like this, because Gwen ignored Hanne. She simply unlocked the boot and pulled out her suitcase, without meeting the gaze of her rival.

But before the dust from the Manta had settled, she dropped her suitcases, sat on the ground, and holding back tears said, "Who the hell is *she*?"

She said it quietly, in a tragic way that made me feel like putting an end to it all.

"Did you have her . . . waiting in reserve?"

"She moved in of her own accord," I said, kicking at the gravel.
"I see."

I wondered how it would look inside, if Hanne had scattered her clothes and books everywhere.

"When was it over between you two?" Gwen said. "Just now?"

"Long before. She got an idea into her head. I only asked her to check in on the farm."

I looked towards the potato fields. I wanted to run up and see if all was well, or if there were signs of blight and potato scab after all.

"And when did you ask her to leave?" Gwen said.

"When I realised she'd somehow got it into her head that we were going to be together."

In the summer heat a faint scent of tar emanated from the timber walls.

"You fucking bastard . . . When was that? Just before we left Shetland?"

"No. Before you said you were going to Edinburgh. Before your secret mission to Lerwick."

"I thought we were done with all that. And when did you last sleep with her? Answer me honestly."

I squirmed. I cursed. I kicked at the gravel again. "A few days before I left for Shetland. But it was the first time in—"

"I see, *now* I understand you," she screamed, and stood up. "This is my punishment for having gone behind your back and sneaking into the lock-up!"

"Now stop it!" I said and tried to grab her arm.

"You bastard!" she snapped, pulling free. "Do me one favour, drive me to the train station. And that'll be the last we see of each other."

"I chose *you*, Gwen. That time I called her, I wanted to be with you."

"So answer me one thing. If you had come home alone and she met you like that. In that dress, probably with no knickers and horny as hell. Would you have resisted?"

"She didn't know you were with me, Gwen."

"So answer me, then! And you're saying that she *didn't know about me*? Are you such a damned coward?"

"I—"

She threw the suitcases into the boot. "Back home only sluts dress like that. Jesus. Wish *I* had her tits. I bet you fantasised about her while we were having sex."

"Now that's enough, Gwen!"

She put her hands on her hips and walked a few paces away. Looking across the fields, she turned her face to the warm wind which carried the smell of freshly cut grass.

"Damn," she said quietly. "On top of everything I'm jealous of her, walking around dressed like that and she *still* looked elegant. Not a single blemish."

"It's not as if she was *naked*."

"No, as a man you *would* say that, of course. I had it hammered into me by Grandmother: 'You don't do hair, make-up, legs and cleavage at the same time. You just *don't*.'"

She glanced again at the log house, turned and sat down in the only place that might remind her of home. The passenger seat of an English car.

I went inside. The house had just been cleaned. A few of Hanne's books and clothes here and there. On the first floor the double bed was made up with crisp, blue-grey linen. By the telephone was the picture of Mamma and Pappa.

Back outside, Gwen had closed the car door and was staring straight ahead. Grubbe came across the lawn and sat at my feet. The shaggy tail of the forest cat beat slowly against the grass. He looked at me, but he was not going to meet me halfway either.

"Let's go," I said, and turned the ignition. "You can take the train to the airport. I'll pay for your flight to Aberdeen. Or London. Wherever you want. But first do me one favour."

"Excuse me? You don't qualify for favours."

"Just walk through the town with me."

"Why?"

"You'll see."

"Do you mean the tiny . . . the village down by the river? Why in the world would I want to do that?"

"Just from the post office to the shop."

"I don't owe you anything."

"You owe me the chance to explain where I come from."

"I know where you come from. Here."

I checked myself, tried to think through what I had told her about my life, how I had dressed it up. Then I got out of the Bristol, took her by the hand and dragged her across the courtyard, to the barn ramp, where Bestefar's Mercedes stood covered in dust.

"Do you recognise that?" I said.

She put a finger on the boot and wiped off a layer of grime beneath yellow pollen. Rubbed at the windows and looked inside.

"His car," she said. "I saw it in Norwick when Einar was buried."

"We can't avoid this, Gwen. Neither of us. He never told me that he had been to the funeral, or that—"

"Edward," she interrupted. "I'm totally fed up with all this mess. Take me to the station."

"Wait," I said. "I don't know why I'm telling you all this. But you are . . . a part of it. Look at the car door. In the summer someone painted a swastika there because Bestefar fought for the Germans on the Eastern Front. Before I met you, I wasn't how you might imagine. I was a . . . a recluse. My world stopped at these fences. That's how it was until I met you."

"So now you want to parade me around like a trophy? Make people say: Oh, he wasn't what we thought after all?"

"Just come with me, for your sake. It's like . . . it makes sense."

We parked by the post office and soon young boys on bicycles were stopping by the Bristol, trying to get a look at the speedometer.

Gwen had slammed the door, tied the belt around her cape

and crossed the street, where she now stood between the agricultural cooperative and the draper's, fishing out a Craven A. She lit
up and looked at her shoes, smoked as she always had, holding
the cigarette over her shoulder with an open palm. But she was
on guard. A slight twitch, a quick turn of the head.

There was no white Manta to be seen, but from the corner
by the post office the Hafstad boys were staring at her. Indolent
youths hanging out by the sports club noticeboard, gawping at
Gwen and muttering to each other. Mari Øvereng, usually the
busiest person in the village, suddenly seemed to have all the time
in the world.

The village was weaving its thread.

Gwen stood there and I let her, because I could see the defiance building inside her.

"It's nothing," I said when I joined her. "They just want to
know who you are."

"How can they see that from there?"

"They can't. That's why they're staring. They don't mean anything by it, they just want to see what's going on. I'm like that too.
Even I stare, but I've only just realised it now."

"So before you just thought that everyone stared at you?"

"Yes," I said. "I must have."

Something changed inside Gwen. Her movements became
calmer, and she walked down to Nordlien, the grocery where
affluent locals bought steak for the weekend and their children
were able to buy sweets on credit.

Everything stopped when we walked in. She grabbed a basket
and began to fill it.

Would she be staying after all?

She picked up eggs, whole milk, smoked pork belly, black
olives. A tin of Coleman's mustard rather than Idun's. A bottle
of Worcestershire Sauce, and before she put it in the basket, she
looked around as if to say: What's so funny about Worcestershire
Sauce? You *do* sell it.

I had wanted to go into the village to see what it felt like to be someone else. I liked it for a bit, strolling round Saksum with a new lady, wearing Lobbs, buying Worcestershire Sauce and driving a Bristol. But now I felt naked. Saksum told me what neither the cat nor Hanne had said:

Welcome home, but we still don't believe it.

2

SHE STOOD BY THE WINDOW WITH BARE LEGS, WRAPPED up in one of the faded floral sheets that had been in the house all these years. I lay in bed with my eyes half open, wondering what she thought of the sheet. The girl who had worn tailored clothes her entire life.

The sight of her here, with her back to me by the yellowed wooden panelling and my nature photographs. This could vanish so quickly, like a shape slipping away; she was so foreign here that I doubted whether the vision was real, even though she was right there in front of me, filled with an animalistic desire. I dared not close my eyes.

Down at the shop she had said, "Fine, let's go back. Let's keep fighting this eternal battle against some shit that happened a long time ago."

I never quite understood why she changed her mind. Doubt flowed inside me. Perhaps she would undergo this primitive humiliation in the wastelands in exchange for the possibility of finding the inheritance, the sixteen trees of the Somme.

She was an early riser, but I asked myself why she had not put on the kettle for tea. In Shetland it had been the first thing she did. Perhaps it was because Hanne was also in the room, the ghost from yesterday sitting golden-brown on the windowsill, dangling her bare legs.

She noticed that I was awake and turned. And we did what we had always done at this farm, to hold the big questions at bay.

We ate a huge breakfast and said as little as possible.

We smoked on the stone steps as the dew steamed from the grass.

We started up the old tractor and drove it out to the fields.

But she couldn't do it. She wouldn't do any heavy lifting, she just stood there with her feet apart, thinking about the incongruity between what she had to do and the clothes she did not want to get dirty. I showed her how the potato lifter worked and started the tractor. But she had never come into contact with soil in her life; the only thing she really knew how to handle, apart from a tea kettle and a record player, was *Zetland*. Up here she was helpless and apathetic, and she didn't want to ruin her nails.

But in the end she gave it a go. The ploughshare dug down, the earth was opened up and small seed potatoes appeared like jewels. But she plucked them with her fingers not her fists, carefully placing the potatoes one by one in the crate, and it went unbearably slowly, she spent the whole time worrying about her clothes, shaking the dirt off her hands. After about five minutes I looked in the side mirror and realised that she was just standing there, staring.

I climbed down from the old Deutz and leaned against the tyre.

"How do you swear in Norwegian?" she said.

"Sorry?"

"How do you *swear* in Norwegian?"

"Well—"

"I *mean* it. Turn off that bloody tractor."

The cackling of the diesel died.

"How do Norwegians swear? 'Damn' is no good here. Nor is 'fuck'."

A pair of crows circled above us and landed in the spruce trees.

"Well, *faen* is pretty much the universal swear word," I said.

"F-ain?"

" Longer 'a'. You sound like a foreign doctor."

"*Faan!* What else? I need more."

"Try *i svarte helvete. Satan ta. Faen i kølsvarte helvete.*"

She grabbed a clump of earth and threw it at me. "*Fuck satan i svarte kølfaens helvete. FAAN! FAAN I HELVETE!*"

"*Dæven drite* is good too," I said.

She swore away, and suddenly tore open her blouse. Buttons rolled into the furrows. "*Faan!* Mosquitoes! So *itchy!*" She shrieked as she pulled it off and trampled it into the ground. With muddy fingers she scratched at her arm until it went red, then picked up the ruined top and marched towards the farm.

The crows took off and disappeared over the treetops. I sniffed the breeze. The worst that could happen now would be rain. Maybe Yngve would help with the rest. Another thought hovered close by: If I had come without her, there would have been a long row of full potato boxes here this morning, and I'd be resting under the plum tree with Hanne Solvoll.

The crop appeared to be fine, despite weeks without tending. Spring had been warm and the potatoes had been chitted before planting, but it was as though the ground trembled with suspicion at the risk this farmer of Hirifjell had taken.

As I walked down towards the houses I saw a strange figure stomping towards me. She wore a checkered apron, a headscarf and Wellies which were far too big and smacked against her calves, and only when she was thirty metres away was I able to accept that it was Gwen. Twenty metres, fifteen . . . it was both her and it was not. She was wearing Alma's work clothes.

The memory flew at me like an arrow. Alma's voice, the cautious glance out of the corner of her eye. Then the figure was torn away, as if in a sudden gust of wind, and Gwen was there in clothes she could work in. Her make-up had been washed off. She had been crying, and she was still sniffling as she ran her fingers through the earth and scooped up the potatoes. Dirt on her knees, dirt on her fingers, dirt on her mind.

"Why are there ashes in the middle of the field?"

"I burned some furniture," I said after a moment. "I had to do it out here so the sparks wouldn't reach the buildings."

I had been adjusting the potato lifter, and when I turned she was gone. I put down the spanner and caught up with her as she stepped into the circle of charred wood and ashes. The potato vines that surrounded it were taller than the others.

"They grow better here," she said.

"Ash is a good fertiliser."

She pulled up a vine and knocked it against her boot so the dirt trickled off. The pimpernel was bright red and fresh-looking.

"See how nice it is," she said. "Can we have them for dinner?"

"We can take some from another field." I turned to leave.

"But these look so good." She pulled up another vine. "I want these ones."

"Gwen," I said and wiped my hands on my trousers. "It wasn't furniture I burned here – it was a coffin."

She stood with the potato vine in her fist as I told her about the flame-birch and Bestefar's second funeral. She was quiet for a long time, and began to pick the potatoes off the plants.

"Why couldn't you just say so," she said. You're one to talk, I thought. "But thank you for letting me know. I haven't changed my mind. In fact, I'd like to eat *these* potatoes in particular."

She put them into the pocket of her apron and again stepped inside the blackened circle, kicking at the charcoal of the flame-birch. Like Alma dancing on Bestefar's grave.

"There's something here," she said. "The blade of a knife."

I picked up Bestefar's blackened bayonet. On the tang was a stamp which had been hidden by the flame-birch stock. With my thumb I rubbed away the soot, to reveal a number and a swastika.

Why had I thought it was a Russian bayonet? Was it something said an eternity ago, to satisfy a child's questions?

I recognised the number, the same as on his Mauser, still hidden under the insulation in the attic. Was it *his* broken bayonet? Had he broken it himself, during or after the war, in anger at what he had seen?

That was our story, over and over. I expected the truth, but found only the ashes of truth. I stood alone, to judge fragments from the past.

We worked and worked. Gwen gave in to the grind, slept with tender muscles and awoke hungry. She would never be a proper farmer, we both realised that, but her get-up-and-go flourished and turned us into a team that worked well and quickly.

The lorry from Strand Brenneri collected the tonnes of seed potatoes sorted into large wooden boxes. The driver nodded and commented on the quality of the crop, offered his condolences and cast a surprised look at Gwen.

I watched the lorry rumble across the cattle grid and turn onto the county road. The drone of the motor receded between the spruce trees.

Autumn stillness.

Before long we would have to bring the sheep down from the mountain. Then winter would come, and I would be tied to the farm. I had been itching to travel to France, but like a slow poisoning, the days grew calmer. Einar's workshop was like a gravestone I passed without thinking, and Haaf Gruney existed in my memory as a place I had left many years ago.

We never mentioned the walnut or Quercus Hall. Instead we drove the Star to the lake in the mountains, cast our nets with the sun glistening on the wet mesh and on the brown-spotted trout. I snapped their necks one by one while she rowed the old wooden boat, which was something she *could* do. She managed the rowing better than either Bestefar or I had, while I took off my sweaty flannel shirt so I stood wearing only a T-shirt for the next catch. I caught myself turning sideways to lift the net, so she could see my biceps. She rowed out to the river mouth, and I watched as she guided the boat through the water, its wake like fine calligraphy on the surface. It was not yet seven o'clock, and I wanted it to be just the two of us, a brand-new start, with no past, no family, born of ourselves.

3

BUT OF COURSE IT COULD NOT LAST. ONE DAY I WAS
heading into the village to shop and had to turn back because
I had forgotten my wallet. As I came into the kitchen I heard
her voice on the first floor. I walked noiselessly up the stairs,
stopped on the stair where I had heard Bestefar's footsteps for the
last time, and listened to her speaking on the telephone, to her
mother, from what I understood. It was then that I knew our time
was coming to an end.

"It's nice and hot down here," she said, and from the rest of the
conversation I could tell that she was pretending to be on a train
to France, but would be home before long.

I tiptoed back downstairs, and from then on Einar, and the
four days of my disappearance, never lost their hold on me. I
began to search the attics for the letters he must have written to
Mamma, to no avail. Settled down with the things I had brought
from Haaf Gruney: the chessboard, the shotgun, the dress. The
newspaper clippings. Looked for some connection. Took the roll
of Orwo NP20 into the darkroom, switched on the red light and
put the film in the enlarger.

Us in the lay-by. Me with my toy dog. Mamma and Einar.

I made prints of all of them. Slowly the images appeared as
the chemicals sloshed about, the contrasts intensifying on the
paper. Features emerged, and I could see how happy Einar was,
standing there next to Mamma. In another photo, the boy who
would grow up to be someone the dead could rely on.

Afterwards I developed my own films from Shetland.

Bestefar's face on the first frame, then Haaf Gruney, a few of Gwen. Finally, a detailed image: the war map from Quercus Hall, creased and dog-eared, frayed by wind, sodden with rain and dried again.

I made a large print. My hands shook a little as the map of the Daireaux woods appeared in the developing tray. I could see the ponds beyond the river.

Outside I heard Gwen call: "Where are you?"

"In here!" I said.

"Can I come in?"

"Give me a minute, I'll be finished soon. I have light-sensitive paper in here."

I switched off the red light. The war map was still imprinted on my retina. I hung the prints to dry under the bench and came out with only one, of Gwen rowing *Patna*.

"Hm," she said and studied the photograph for a long time. "Is that how I look?"

"You did then, at least."

"I'd be happy with that," she said. "Being like that."

"Come on," I said. "Come outside with me. I'm going to show you something."

She put on her coat and we went up to the flame-birch woods. We had walked past the previous day, but now the dew had taken hold of the twigs protruding from the ground and hundreds of delicate spiderwebs shone white in the sun.

"Do you see?" I said.

She was amazed. Shook her head slowly.

"They weren't here yesterday," she said. "Or we didn't see them. It's like an entire world, an invisible world which has become visible only now."

She crouched down by a stump, carefully touched a spiderweb.

"Are you going to take a photo?" she said, coming back to me.

"No. I don't need to. I'll remember this."

*

A name from the newspaper clippings had stayed in my head. J. Berlet. The policewoman from the investigation in 1971. I imagined myself in France, going to the police station and asking for her, then driving to a private address written on a slip of paper.

No, I told myself. It would be nothing more than a wild goose chase: bad French and misunderstandings, endless driving on three-lane motorways, reticent locals. Much better to call upon an old ally.

I went out to the log house and picked up the telephone.

"Televerket's international enquiry service, how may I be of assistance."

"I need to get hold of Regine Anderson."

"Which country, which address?"

"No, it's not like that. She works for Televerket."

A little later I had a number with a 33 prefix for Jocelyne Berlet. Regine Anderson had told me in a whisper that she was now living in Péronne, not far from Authuille. That same evening I dialled the number from the log house.

"My name is Edouard Daireaux," I said in French. "I'm calling from Norway about a police case from twenty years ago, a search for a boy in 1971."

She did not answer and I could hear her shifting her grip on the receiver. I reminded myself to use the polite form.

"Is this the Madame Berlet who investigated the disappearance?" I said. "I was the boy who—"

"Edouard?" she said suddenly. Her voice sounded maternal.

"Yes."

We were silent for a long time.

"In 1971—" she began, and then stopped herself. "No. As a matter of fact, this is highly inappropriate."

"I was wondering whether I could meet you," I said.

"I'm sorry," she said rather sharply. "I no longer work for the police. I retired in 1975."

"Yes, but—"

"Tell me, what is it that you actually want?"

311

"If I were to come to France," I said, "would you meet me and tell me what happened back then?"

"The case was never solved; it would be against regulations to talk to you."

I was stretching the boundaries of what I could say in this foreign and yet strangely familiar language: "Perhaps, but will you talk to me anyway?" I hoped I did not sound too abrupt.

Jocelyne Berlet cleared her throat. "When were you thinking of coming?"

This is it, I thought. Now I'll settle it. I fix a date, and then I'll set out.

"Quite soon. But how well do you remember the case?"

"Remember? I remember it as though it happened . . . well, not yesterday, but last week anyway."

4

AN EARLY FROST.

A thin layer of rime in the courtyard as I walked across to fire up the griddle in the baking cellar. The thinly chopped aspen burned quickly and fiercely. Baking *lefse* today. She enjoyed the simple life here, the little rewards for everything we did with our own hands. Like on Unst, where the effort of taking a bus to Lerwick in the pouring rain was repaid with a new album and an evening spent on a comfortable sofa drinking sugary tea. Brief periods of work and instant gratification, without the whole world gawping from the stands.

All of a sudden I felt afraid of a time when I might be alone, standing at this rusted griddle, thinking back on the strange period when someone had been here with me.

I went through the side door and recognised a familiar smell. It was the room with the deep freezer, the one we used for meat. I pulled open the lid a crack, and from it came a waft so disgusting that I imagined I could see it, bursting with decay, heavy as burning oil. The room was in darkness, and when I flipped the switch, nothing happened. The fuse must have blown.

It was there, in the stench, that something struck me.

The letters, Einar's letters to Mamma. Where would you hide them here on the farm? Somewhere that was accessible. A freezer. I fetched a sack trolley and pulled it outside, wedged open the lid to let out the air.

Gwen came across the courtyard with a bucketful of *lefse*

dough and some almond potatoes. "What is that horrendous stench?" She stopped and grabbed her nose.

"Spoiled meat."

She stood with the bucket. "Now I know what he meant."

"Who?"

"Grandfather. He told me that hundreds of soldiers fell in front of the trenches. They hung there on the barbed wire and rotted in the sun. He said the smell was like that coming from a broken freezer. But we never had a broken freezer so I didn't know what he meant. Not until now."

"Do you still want to fry some *lefse*?" I said. "Or should we forget it?"

She looked at me. "I want to. More than anything, I want to fry *lefse*. Just put that lid down."

While I rolled out the dough, I imagined how I would have hidden the letters. I might have wrapped them in plastic and tied it with twine, then put flat stones around it to get the right weight and hardness of frozen meat before wrapping it again. Marked it "Elk heart – 1967" or similar.

Later that day, after Gwen had declared that she had never tasted anything better than fresh potato *lefse* with goat's cheese, I went back to the freezer. I slit open the packages on top, but found only minced elk meat which I threw in a wheelbarrow and wheeled to the edge of the woods to bury it. Three trips later I got to the bottom. Bloody water had pooled in one corner. The stench was overwhelming.

What foolishness, I said to myself. Bestefar would have thrown out any old meat. I found a package with my writing on yellowish-brown freezer tape. "Mallard – 1981". I tore off the *Lillehammer Observer* from August 20 of that year, the day the hunting season opened. The first time I had been alone in Laugen with Pappa's 16-calibre.

There were no letters in the freezer. Of course there weren't. I had pictured it so clearly in my head, me surrounded by the nauseating odour of decay, cutting open packages as blood-pink water seeped through onto the paper, recognising Einar's writing and hurrying because the words were blurring, disappearing as I watched, a life story trickling away in an interplay of blood and ink.

But these were just imaginings.

I now felt embarrassed to have read her letters. Had I learned so little from them? Where did she prefer to sit when she wrote to her father? Where on Hirifjell did she feel closest to him, and away from the others?

The answer was right in front of me.

I found Einar's letters hidden between the materials in the cabinetmaking workshop. Planks of wood that neither Sverre nor Alma would ever touch, left to be the indulgence of the prodigal son and heir. It was as if they said: Come here and look! I may be no good at farm work, but look what I can do with a chisel and a plane, with lacquer and linseed oil.

There they were, safe from mice and other vermin and prying eyes, pressed flat and dry, and with a little sawdust on top. Twenty or so letters in which Einar explained what had happened in 1943 and thereafter. He had written in French, with even margins and on both sides. Just as they had begun to fill out both sides of *my* blank page, because now I understood Mamma's responses to him.

She had been hateful when Einar showed up at the farm in 1967. Perhaps she had suspected that his plan during the war had been to get his hands on the walnut. And it emerged from Einar's letters that he, by asking some remarkably detailed questions, had needed proof that Mamma was in fact who she claimed to be.

I recognised myself in both his and Mamma's manner of speaking. I finished reading the letters and straightened up, feeling the same certainty that Mamma and Einar had been blessed with.

Einar had been my grandfather.

I fiddled with the cabinetmaking tools, looked at his proud handwriting. It was as though he were standing in front of me, opening a direct line to the year 1943.

5

WHEN EINAR FIRST ARRIVED IN AUTHUILLE, HE HAD
been gruffly rejected by my great-grandfather. Einar had told
Edouard Daireaux about his mission, about the sapper's map, and
had handed him an envelope from Winterfinch with the final
amount due for the timber.

Edouard had asked him to leave. "The money is as agreed,"
he said. "But that's the least of it. The *actual* deal was for the
woods to be cleared of shells, but instead he's set up a minefield
in there. When the timber goes, it'll be impossible to get anyone
to clear them. You should get out before the Germans see your
documents. And keep the money. I don't want a nitwit with a fake
name in my woods."

Einar slowly began to realise how meaningless his task was.
On his journey through the country he had run into German
patrols and witnessed the emaciated population, but what of
the *effects* of the war? He had witnessed little of that. Only now
did he understand that the country he so admired had become
impoverished, enveloped in a sadistic darkness.

He accepted the envelope, went on his way and contemplated
how far from reality he had been. The flight to Shetland was provoked by a concern for his brother, for as he wrote: *I knew that
once I had left Hirifjell, Sverre would not enlist with the Germans
again. He had a farm to run. And I had no intention of becoming an
allied soldier. I just wanted to get away. Until that point, whenever I
heard the word "German", I just pictured Sverre in uniform.*

When his skills as a boatbuilder had become superfluous, he

had met Winterfinch, the first man in many years with whom he could discuss fine cabinetmaking and woodwork. Einar made several pieces of furniture for Quercus Hall, and when he completed an armchair with an ebony frame, Winterfinch began to call him by his first name. *Then one evening he told me about a consignment of walnut trees that were of great sentimental value to him.*

Winterfinch had bitterly regretted not felling the trees before the war. From 1941 onwards he became more and more desperate, having heard reports that the Germans were gradually moving further into the country and felling timber – whether the landowners allowed it or not – for the massive fortifications along the Atlantic coast. The agreement was that Einar was to fell the trees, hide the shipment wherever he thought best, and await further instructions.

Now in Authuille, and faced with the reality, all those plans were discarded. His life was meaningless, he would be useless as a soldier, and the only thing he could do well – decorative cabinetmaking – was not exactly sought after in a war. But he had not gone far before he heard the crunching of gravel behind him. A scrawny girl was following him on a bicycle. She wore worn work clothes and a kerchief over her blue-black curly hair. Bony and bedraggled, like many others at the time, but with the gaze of a bird of prey and fleet movements. She was the farmer's daughter, Isabelle Daireaux. He had seen her at the house while he had been talking to Edouard.

"Are you a good Frenchman?" was the first thing she said.

He wanted to answer that he was as good a Frenchman as a Norwegian could be, but instead he simply nodded.

"Do you have the sapper's map?" she said.

Einar stared at her.

"I've heard told that the resistance movement needs explosives," she said. "And the shells from 1916 are intact."

"But they're extremely dangerous," Einar said.

"That's why they're needed," Isabelle said.

"Have you just come up with this idea now?"

"No. On the bike ride here."

"That's not a long trip."

"I'm a quick thinker."

"And your father?"

"He's too cautious," Isabelle said, and led him off the road. Einar cut open the lining of his jacket to show her the sapper's map, which had been hidden there for his entire journey as Oscar Ribaut.

"The worst thing about hearing shots," she said, "is when there's only one. They murdered my schoolmistress the day before yesterday. Buried the baker alive because he was in the resistance. Our crops were confiscated even before we got them back to the house. We're fattening up soldiers who are shooting our countrymen."

She asked Einar if he had noticed the girl who stood by the henhouse on the farm. "That's my sister," she said. "How old do you think she is?"

Einar shook his head. "Twelve?"

"She's fifteen," she said. "She's not growing properly because we are starving. Even though we live on a farm."

Isabelle had persuaded her father to let Einar stay, in exchange for his travel money from Winterfinch and helping out with the farm work. Later they had taken the map and gone carefully into the woods. The ground was pocked with craters from shell strikes. In some places young conifers sprouted between charred, broken trunks, in others the forest floor had simply lost all life. Here and there, hidden by undergrowth and dead grass, they could see how rusted artillery shells had been positioned in a perimeter to prevent illegal felling.

Within this area of fire and destruction stood the walnut trees. Disfigured and charred, on land broken up by the digging of the sappers. The trunks were so enormous that they could not reach all the way around them. Small branches, like stunted arms, had developed, but the foliage was yellow and wilting. A peculiar, dead smell hung in the silence.

Isabelle had never been into the woods before, and she shared her parents' disdain for Winterfinch's indeterminate plans to clear them. The woods had been valuable before the war, as building material and as firewood, not to mention the rich crop of plump walnuts. The ancient trees had seen both Napoleonic wars and revolutions before the gas bombardments had made them die out.

Einar just stood there imagining how this place must have been before the first war, a spacious, peaceful parcel of land with lush green trees that touched their neighbours. It resembled his own woods in Hirifjell, apart from the fact that back home it was *he* who had inflicted the wounds on the trees. Then Einar had noticed something by its absence. Birdsong. The place was absolutely still.

The Germans had probably suspected that someone might collect the explosives in this way. The Daireaux woods were surrounded by patrolled roads, but they were situated on a slope that led down towards the Ancre, and with the help of the map they could sneak unseen through the thicket and follow the secure path to the walnut trees. When they got there, they found everything neatly arranged; behind the perimeter the sappers had gathered shells in large mounds, and had even removed the detonators.

In silence they took the shells apart, removed the T.N.T. and placed it near the water. Einar had no idea how they were transported from there, but by the following day they were always gone.

In the woods they kept stumbling upon scrap metal from the previous war: tin plates, helmets, rifle casings, military boots with the remains of a leg inside. That this place was a mass grave did not appear to affect the two particularly. But Einar, who had come from the barren and treeless Shetland Islands, could not spend long near a tree without feeling the cabinetmaker in him come to life. One day, when he was alone, he dug down to the roots of one of the trees and felled it with an axe. Only when the tree was toppling did it occur to him that the impact could set off explosions. He threw himself to the ground, the earth shook, but no explosions came.

With a rack saw he cut out a block of the wood and moistened it. Immediately it brought to life an extraordinary interplay of colours. Powerful black rings on a reddish-orange background which appeared to glow. From his purchases for Ruhlmann's he recognised the quality grades of walnut; it did not take him long to calculate that, in peacetime, the timber would be worth an enormous fortune.

Isabelle had been furious that he was so distracted from the task of collecting shells.

One night she appeared at his shack holding something behind her back. An old burial cross, its paint peeling, still damp and blackened at the base.

"You don't want to know where this came from," she said, placing the cross on the floor. "Just make two more which look the same. And carve in these names." She handed him a piece of paper.

The names were Jewish. The years of death were 1938 and 1939.

"But Jews don't use crosses," Einar said.

"Please, just do it," Isabelle said. "Prove that you really are a cabinetmaker."

By the light of a candle he took two old planks, inserted a crossbeam and carved in the names with a chisel. He splashed on some white paint and rubbed the crosses with dirt and old engine oil to give the impression of age. That night they went back out, she taking him by the hand now and then to guide him through the fields. There was bombing in the vicinity that night, and they heard shots not far away as they slipped towards the churchyard. They took crosses from two graves and replaced them with fake ones, patting the earth around the base.

As they sneaked back she told him that the Germans had arrested a Jewish family. Since the occupying forces traced Jewish ancestry back two generations, the fake crosses would carry the names of the family's grandparents. The next day a priest would lead the Germans to the churchyard and point to the crosses to show that the family had been buried according to Christian tradition.

Einar wrote about this incident so that Mamma could verify his story. *You can visit these families yourself and have the story confirmed to you*, and the address of a man by the name of Staniszewski was also written there. When his letter came to the circumstances of the arrest, his writing grew more shaky, and I read a sentence which sent shivers through me:

Isabelle owned a pretty blue dress she had made before the war. A few days before the Germans came she had worn it, and it was a sight which would later haunt me each night.

For when the Daireaux family was taken, Einar had not fled straight away. He was suspected of having informed on them by others in the resistance movement, yet still he recklessly went to the Daireaux farm to obtain some memento of Isabelle. All the farm sheds had been burned to the ground, the Gestapo's means of driving out any children who might be hiding. Everything had been razed, and their dog lay in a pool of dried blood. Inside the barn he found the blue summer dress, trodden into the hay and filth.

I took it as a memento and as proof, he wrote, *that I was no informer. But I was in mortal danger and had to flee. I had a friend in the area, a skilled cabinetmaker who had worked with me at Ruhlmann's workshop in Paris. His name was Charles Bonsergent, he came from a fishing village a day's journey away, and had, like me, moved home with the onset of war.*

Einar had remained in hiding with Bonsergent until D-Day. He discovered that Isabelle was in Ravensbrück, and when the telephone lines were re-opened, he contacted Winterfinch and asked him for money for travel and pay-offs so that he could bring her home as soon as Berlin fell.

Winterfinch refused, simply asked where his timber was.

And that was when Einar took the walnut wood hostage. He and Charles Bonsergent travelled back to Authuille, where they felled the remaining trees, dug up the roots, and transported everything to a safe hiding place. He considered the walnut to be Daireaux family property, and refused all of Winterfinch's subsequent offers. Only Isabelle could decide the price, he said – or, later, Isabelle's child.

In later letters Einar mentioned that the inheritance was in the same place, but without saying where, *for reasons we both know well – rumours travel fast on Shetland*. It seemed that it might be in France.

Mamma had visited him out on Haaf Gruney. He must have given her Isabelle's dress there, for in a letter he wrote *How good it is to see it filled with life again*. Einar insisted that the inheritance should go to Mamma, and that he would help her assert her claim to it. But in the first years after I was born she was unwavering. It was not until I read the second to last letter from Einar, written in the summer of 1971, that I saw their plan in black and white:

I have met with Mr Winterfinch. It's the first time we have spoken in a long while. We have glared at each other now and again, I must say, because he has been looking for the inheritance all these years. He even bought a summer house near Authuille. I've felt sorry for him, I really have, not least when I've seen him on Unst with his grandchild, a little girl. But it would be wrong if he simply took the goods, even though I grant he has some right to them. He once said, "Why clear the woods of shells when the family no longer exists?" Since there is now an heir, I told him, the most suitable arrangement would be to give her enough money to buy her family farm back. "She is an imposter," was his response. Ugh, this whole affair has become an evil assortment of events that ought never to have taken place. Your mother will never be returned to us. But now we can have the matter settled, and then I will move away from here. I leave it to you, Nicole, to choose what you think best.

Winterfinch spends September in the Somme every year. The closest hotel is in the nearby city of Albert. I forget what it's called now, something to do with a basilisk? I am staying at a cheap guest house on the outskirts, I cannot bring myself to stay near Authuille itself. There is a good restaurant there, by the way. If you and Winterfinch were to dine together, you could hear his story and would be able to discuss a fair price. Yes, it's worth a fortune, a proper fortune. But do what you want. If you come to an agreement, you could either tell him where the wood is, or let me know how to deliver it. Remember that he can be volatile and difficult; he changes his mind constantly.

Listen, let's talk through the details on the telephone. It's good of Sverre to lend you the car. Don't tell him I've written.

Another thing: if you two want some time alone, I wondered whether I could look after Edvard? I've never looked after a child, but I know him better now after the last time. I've made him a little dog out of beech wood. It can wiggle its ears and shake its tail. When I finished it I just sat there, lost in thought. After all these years of working with wood, it's the first toy I've ever made.

My very best wishes to you, and pass on my greetings to Walter. And yes, do call me. The telephone box by the ferry terminal. Sunday as usual. Six o'clock, also as usual. I'll be there. As usual.

E. H.

6

GWEN SAT ON THE STOREHOUSE STEPS WITH GRUBBE.
He was lying on his back in her lap, curled up, her fingers in his
long fur.

"He smells so good," she said, burying her nose in his stomach.

"That's where he can't reach to clean himself."

"Mmmm, lovely. I've never had a pet, not even a dog. Even
though we had so much space."

I sat beside her on the stone steps. It was getting colder each
day. "Autumn is here," she said. "When do we bring the sheep
down from the mountain?"

I did not answer straight away. And perhaps she had got to
know me too well, could recognise the look which suggested a difficult conversation was on its way. Why bring up the body when
the flowers on the grave are in bloom?

"Oh well," she sighed. "I knew this day had to come. You want
to go to France, don't you?"

"Is it that obvious?"

"Your hesitation. The way you walk. You're distracted, you
don't listen to half the things I say."

Grubbe must have sensed the unease in her, for he leaped off
her lap.

"You want to go now," she said, and plucked a hair from her
sleeve, "because when the sheep are back down, you can't leave, is
that it?"

"I found letters," I said. "An exchange between my mother and
Einar. In French. You can read them."

"Why would I want to read their letters?"

"It says that they went to France in 1971, to meet your grandfather."

She got up and walked a couple of steps away. "I see, so that's it. If we go to France together, it's over between us."

She knows something, I said to myself. She's trying to keep something secret.

"Why's that?" I said.

Gwen picked up a stick and threw it into the nettles.

"Are you so stupid? Isn't it obvious that Grandfather had something to do with their accident? And it won't be a pleasant discovery."

I was going to say that I knew about their summer house, but instead I asked, "So you've never been to Authuille?"

"Me? No, never."

I shrugged, and she began to lose her temper.

"What is it you're trying to say? I told you, we'd drive down through Europe every summer in the Bentley. But it was always a straight line to Dover, across to Calais, and then on to Paris. He even accelerated when we passed the turn-off for the Somme."

"I see."

"Don't you believe me? Well, man up and decide, Edward, do you want me or not? You don't trust me, do you. You think I know something about your disappearance."

"But you *do* know more."

"No, I told you! And I don't *want* to know more either."

It struck me that, before the summer, Hanne had said something similar. On this very spot. And I had left her because of it.

7

A FEW DAYS LATER WE CROSSED THE FRENCH BORDER. It had not been an especially cheerful journey; it felt as though there were two strangers sitting quietly in the back seat listening to what we said.

But Gwen had climbed in with her worn leather suitcases after the Bristol had gone for two days to the workshop at the Mobil garage in the neighbouring village, where it was treated like an old friend, oiled and tuned up, and given new tyres to replace the cracked Dunlops Einar had been driving on. The rattling and roaring of the Bristol had disappeared, and our chatter became more audible.

Then we raced down the country, took the ferry to Kiel and stopped overnight at a hotel in Belgium.

I had not brought the Leica, but the prints from 1971 were hidden in my suitcase, along with the war map. Perhaps she had something similar up her sleeve. A key to a holiday house, for example.

That night we slept together, and in the morning we chatted like before, but bitter words were beginning to creep in at the slightest disagreement.

Towards late morning we were there. The flat, rolling terrain around the Somme river. The road passed endless acres where power lines curtsied between the high-voltage pylons. It was the landscape of oblivion, a landscape to drive past, the same in every direction, a dusty haze of heat with no fixed point for memories.

A French radio station screeched through the car speakers.

They spoke too quickly for me to be able to follow, but Gwen chuckled at their jokes, hummed along to the showy pop music even though we agreed it was terrible.

The radio talk unsettled me. I was on my way to my family domain. To my mother's country. I began to sense an expectation that I should belong, that I should feel at home. We stopped at a kiosk but I lost the conviction I had had when I spoke to Jocelyne Berlet, and I began to stammer. Everything sounded wrong now that I stood on French soil.

I pulled myself together and bought some Gauloises, the blue soft-pack with its Gallic helmet. The French, it seemed, did not make packs of twenty but of nineteen, so the three rows of cigarettes rested in each other's arches. I found the distinction pleasing.

Have a cigarette, Edouard Daireaux, I said to myself.

Some miles later Gwen placed the Michelin map on her lap and said, "Take the next turn."

From there on it was serious. A sign pointed towards Albert, the town nearest to Authuille. We had not booked a hotel, had no plan. A haze shrouded the autumn sun, the landscape was flat, the tractors were ants on the horizon.

Then the war took over.

Not Einar and Isabelle's war, but Duncan Winterfinch's. We passed a small cemetery with white gravestones. They were dense and symmetrical, like a troop of soldiers on parade. I said this to Gwen.

"They *are* soldiers," she answered quietly.

Further on there was another burial ground, then one more. There were no cars behind us and I braked on the crest of a hill. From there we could see four large cemeteries at once.

At the same time it was as though the noise of battle from my own history had begun to rumble. We were approaching the front, the place from which I had disappeared, the gas shells waiting to go off, the pool in which they drowned. The Bristol hummed quietly and faithfully, ready to lead me there. But I could not do it, I needed something in which to hide my own past. So I stopped at a cemetery and told Gwen that I wanted to look around.

A few hundred gravestones. They still had visitors after so many years. In several places there were letters amongst the flowers and I crouched by one that was laminated in plastic. Written recently by the granddaughter of a private who died on July 1, 1916, she told the soldier that her grandmother *missed you terribly, never remarried and took great delight in bringing up her only child*.

Gwen sat in the car, staring blankly forward. I walked in a wide loop, came back towards her, but then turned and headed for the cemetery a couple of hundred metres away. It was twice as large, and almost all those buried there had died on July 1, the first day of the battle.

We drove on, and to escape the silence I stopped at a burial ground on the outskirts of Albert and once more walked alone within the metal enclosure.

I felt dizzy. It was like standing at the foot of a tower and looking up, or on a ferry looking down. The crosses stretched as far as the eye could see. The names were French. I walked to the far end and tried to estimate the number. I got to three thousand, and when I turned, three thousand new names faced me. The graves were back to back. The French graves were rarely decorated. Half of my blood is that of a people who allow the past to be the past, I said to myself. The Brits, on the other hand, adorn their graves; the Brits are the ones who refuse to forget.

"In the whole of France," Gwen said when we got going again, "there are 930 hectares of cemeteries from this war. Are we going to visit them all?"

Just then, as though we had been pulled there by the current, there was a sign for Authuille.

"You've driven past," she said flatly.

"It's too . . . heavy," I said. "I almost feel as though I want to go home."

"Look at that," she said, nodding towards the highest point in the landscape, at the head of a gentle valley.

There stood a strange colossus, its colours blending into the mist. A large arch which broke with the surrounding landscape, an immense statement.

"Thiepval," she said. "The memorial to the missing."

I stared at her.

"That's enough!" she snapped. "I have *not* been here before. But I paid attention in history lessons."

"Don't get so worked up! The gunmaker told us about Thiepval. It's just that it's so enormous."

"Do you know *why* it's so enormous?"

I put the car into neutral. "Tell me," I said.

"To have enough space for all the names. They're carved into the pillars. Seventy-three thousand soldiers. And these are just the British soldiers who were never found or couldn't be identified. That's where you'll find most of the soldiers from Grandfather's company. And no, I don't want to go there."

A tourist bus came up behind us and I drove the Bristol onto the grass verge to let it pass. I looked at her expectantly, but she said nothing more.

We drove into a gentle river valley and followed the road to a viewing point, where we got out. Through the mist I could see yellow autumnal woods at the bottom of the slope.

So here I was. I had formed so many pictures, imagined the crystal-clear Ancre flowing over rounded stones, that the woods were a dark nightmare of thickets and black stumps that stretched for kilometres. But in reality this was a broad quagmire, even the river was no larger than a flooded creek. Small groves of trees. The bottom of the valley a mass of stagnant pools and channels of greyish-brown water. And I saw the war as it was. Hundreds of thousands fighting in a wide-open landscape. No mountains, no hills, the advances visible from afar.

Finding the Daireaux woods ought to be easy enough. It was

just a matter of finding a moment alone, pulling out Winter-finch's map and matching it with the surrounding terrain. Their machine guns had been where the walnut trees stood. Not far from Speyside Avenue, the supply line where he was found with his arm torn off.

Gwen eventually got out of the car and leaned against the bonnet, her hands on the warmth from the radiator.

"Are you yourself again?" she said.

I shook my head. "Not really, no. Somewhere down there are the woods." I pointed down the slope.

"And the farm, where might that be?" she said.

I had imagined it being in the vicinity of the forest grove, but the bottom of the valley was so boggy that it could not have been farmed. Dotted about were some caravans, though I could not grasp how people could live in this waterlogged area.

"It's hard to say," I said. "It must be up on one of the hillsides where the soil is better."

Gwen wore a light-blue, short-sleeved shirt. On her wrist ticked her grandfather's watch.

"Do you want to see the woods?" I said. "The place where he lost his arm?"

She straightened her clothes. "Yes. I've decided that I do. Anyway, we'd only see them from a distance."

My silence gave me away. I knew I would have to force myself to relive everything, place by place and at the right time of day, and see what came to me.

"You're not intending to go *into* the woods?" she said.

"I don't know yet. But I have to be there early in the morning. At the same time it happened."

"There's no way you're going into those woods," she said. "If you do, I'm going home."

I looked down into the landscape in the hope of recognising some sign from nature, a series of *déjà vus* which might make the events of 1971 come to life. But there was only Gwen.

We did not have to drive far before coming to a sign with

the place name that had followed me through life like a troubled spirit.

Authuille.

"Come on, let's just to do it," I said. "But we'll go on foot."

We got out of the Bristol near a graveyard and wandered down towards a cluster of houses. A Renault 4 passed us at full speed, my lapels flapped in the slipstream. Some mutts were barking nearby. Many times I had visualised how Authuille would look, a bustling town full of life, with clothes lines and people looking out of their windows, but it was a quiet village with no shops, just rust-red, dusty brick houses and little old cars. Two boys played football on a gravel drive. We walked past a small garden, a woman was pulling up weeds but did not notice us.

Gwen took my hand. Her grip was not firm, but it was as though every nerve fibre ran into mine. And yet everything that was pure and genuine seemed to flow in only one direction, nothing flowed from me to her. I hated feeling so suspicious, while she squeezed my hand as though we were walking down the aisle.

Then she let go and suddenly we were out of Authuille. Another infinity of fields sprawled before us. We walked back to the graveyard and searched amongst the headstones for the name Daireaux, but found nothing. It was as though my parents' story was not true. And yet I was the living proof.

"There's a river down there. Should we follow this?" Gwen said and pointed to a road which went steeply downhill.

Now it was as if another Authuille appeared, the Authuille from my memories. Because a hundred metres further down was a building I was certain I had seen before. Auberge de la Vallée d'Ancre. A beautiful restaurant, reddish-brown brick walls with latitudinal white stripes.

Out of the haze of my memory, something was emerging.

"Why are we stopping?" Gwen said.

"It's here."

The feeling I had hoped for, it was right here. Like something subterranean had moved and was crudely digging around, searching for a point of exit.

"What's here?"

"I've been here before."

"Do you mean that? Do you actually *remember*?"

They came in a flash, a few frozen images seconds apart. Pappa holding me under the arms and lifting me up, Mamma saying something to us. They were speaking excitedly, a word was constantly repeated.

And then I made contact, Pappa's voice was clear and pure in my memory.

"Perch," I said.

"What?"

"Perch. Down there, Mamma and Pappa said something about perch."

Gwen set off down the hill.

"Wait," I said. Another image was forming, but it was woolly and strange. "I remember something brown and white," I muttered, "and maybe some letters."

We were still a little way from the restaurant. As if in a trance I stood and stared at the brick building below us. At the grass along the wall. At the narrow staircase.

"So you remember perch and something brown and white?" Gwen said.

"I do," I said. *I do*, as if I were standing in front of a priest.

We bounded down the hill. "Look!" she said. "On the wall there's a menu."

It was in a glass frame with a rust-brown metal edge, the word MENU in white letters. A child would have to be lifted up to see it.

"There you go," Gwen said. "*Perche en sauce safrandée*. Perch in saffron."

Pappa, who hardly knew any French, must have held me up to see the menu. Then Mamma must have joined us and translated.

A recognition, a hazy memory which left a churning in my

gut. I crouched down to the height of a child, noticed the smell of the grass in my nostrils, heard the rippling of the river change tone at this height, and my memory fell into place. The grass was different here, flowers we did not have back home, and deep in the mist of the past I thought I could hear the echoing voices of Mamma and Pappa.

Perch.

The word ricocheted in my memory, bouncing off the walls of a mildewed cellar, and I heard Pappa's voice again. Not clearly, just the ring of it.

And then I had it at last, a perpetual gift, a genuine memory of Pappa.

We stood in the cottage in Hirifjell looking at the names of fish on a wallchart. The memory solidified, and then we drifted away from the farm, back here again, where I felt his hands holding me under the armpits and together we said *perch*.

"It's still on the menu," Gwen said. "It must be their speciality."

The door was locked. I stood outside trembling. Before long a small lady in her fifties appeared and opened the door a crack.

"*Nous n'avons pas encore ouvert. Avez-vous réservé?*"

She was so brisk that I did not grasp what she said, and only a few seconds passed before her patience ran out and she closed the door.

"They've stopped taking lunch reservations," Gwen translated and looked at her watch. "Asked if we had reserved a table for this evening."

I said nothing, just held on to the image in my memory.

"What's going on?" Gwen said. "You seem so . . . happy?"

"It's the first time I've remembered anything proper about my father. His voice in my ear. He was unshaven and scratched my cheek so much that it hurt. My mother was excited about something. They seemed so . . . complete."

She took a step towards me and smiled. She must have hoped that this would be enough for me.

334

"One thing," I said. "The lady said something about reserving a table. They'd write that in a book, wouldn't they?"

"Yes, of course."

"Maybe they keep the books?"

In a gallant demonstration of charm and public-school French, and with a few francs slipped under a crocheted tablecloth, Gwen managed to persuade the lady to fetch a worn, light-blue book. She wrinkled her nose and blew the dust from the cover out of a window.

"Perch in saffron," Gwen said. "Have you had it on the menu for a long time?"

"Ever since we opened," she said, placing the book in front of us.

On the page dated September 22, 1971, we found it. A reservation in the name of Nicole Daireaux, for three adults and a child.

Pale-yellow tablecloths, the clatter of cutlery being set out. I tried to evoke memories from these sounds, but nothing came.

"Here she is again," Gwen said, dragging me from my thoughts. She had turned the page to the next day. "They'd booked for the following day too. But only for three adults."

I looked at Mamma's name.

"Hm," I said.

"Yes. Who would they have eaten with?"

"It could have been your grandfather. Maybe Einar was going to look after me. But they never came. By then they were already dead."

That evening I ate Mamma and Pappa's last meal.

Perch in saffron. A spice as new to me as it must have been for them. Hints of aroma from the yellow-orange threads which bled into a subtle, light sauce. The scent of the food seeped deep inside me, the flavours like memories themselves, demanding surrender, no intense spices to overpower the senses, just an incomparable, delicate flavour which vibrated in my brain and tuned it to a frequency from 1971.

An indistinct shadow play, diffuse dreams, but I was closer to Mamma and Pappa than ever. I could even feel how lost Mamma must have been that first time in Authuille, when she asked for directions to the farm and was sent away. I sensed what she must have thought when she returned to settle the matter.

I knew that I had to continue, retrace our steps, and go out in the morning light to relive their death.

But I did not let the finality of that thought overwhelm me. Because everything until that moment had been peaceful, and comforting. It had not been a risky, reckless trip. They had been here to conclude something, put something to rest.

A person who is sad does not order perch with saffron.

I hardly noticed how Gwen was slipping away from me. She ate quietly, and halfway through the meal she said something that I would reflect on in the days to come. It was a rebuke, but it didn't sound like one at the time, more like friendly advice. She had looked at my water glass.

"Edward," she said. "When you eat food like this, and when you're given polished glasses like these, you use your napkin to wipe your lips *before* you drink. That way you don't leave a mark."

It was as though she had a premonition that this dinner would be the last we ate together.

We did not make love that night. She pulled the duvet over her and turned her back to me.

Sleep would not come. The rumbling from the trip reverberated in my head, and I got dressed and strolled through the corridors. We had checked in at Hôtel de la Basilique, but it did not feel right to tell Gwen that my parents and I had probably stayed overnight here in 1971.

The hotel had only ten rooms, and as I passed the white-painted doors I wondered which had been ours. They might have a twenty-year-old guestbook here too. I stepped into the cool night air and wandered over to the church. A couple of nightbirds sang in a neighbouring street. A pale-grey Citroën with yellow headlamps rounded a curve and disappeared.

Jocelyne Berlet lived no more than an hour from here. On the drive here I had told Gwen about her. Maybe we could pay her a visit tomorrow, I had said. But what if she told us that a one-armed man had been questioned at the time? I had never felt that I bore a grudge against Duncan Winterfinch, not even here. It was as though something exonerated him. I began to plan the onward trip, towards Le Crotoy, to the doctor's office where I had been found.

Gwen was still sleeping when I came back, and I thought about Shetland, the evening we got drunk in Captain Flint's and checked in at the Solheim Guest House. The snickering, the groping on the stairs before we slipped inside our room, locked the door and undressed each other in the hazy light seeping through the curtains.

The same light surrounded her now, this night, in room 8 in Hôtel de la Basilique, her skin tinted by a yellowish shimmer. I lay on the sofa, slept with the old excitement and the old separation from the first nights on Shetland. The hotel room had become Haaf Gruney and Quercus Hall, with the floor an unsteady sea between us.

I woke to her sobbing. She sat in her bathrobe, holding the photographs I had hidden from her.

"Why didn't you show me these," she said tonelessly, as though talking to the floorboards.

My suitcase was empty. Shirts hung in the wardrobe, trousers were neatly folded on the shelves.

"I photographed his war map in Quercus Hall," I said. "At the time I wasn't sure if . . . I didn't know whether you had other plans."

But she was staring at the picture of me and the toy dog.

"*You're* the one with another plan," she said. "I couldn't care

337

less that you took photographs without my knowing. But you've never shown me these."

"What about you? Can't you just admit that you've been here before, that you have a summer house here?"

She seemed not to hear, she couldn't tear herself away from the photograph. "I don't know what happened to your parents," she said, almost inaudibly.

"Please, just answer me," I said. "Do you have a house here?"

"A house?" she mumbled.

"Einar wrote about it in one of his letters."

"Nonsense." She shook her head. "There was no summer house."

I pulled on some clothes. She did not move. "What is it about that photograph?" I said, buckling my belt.

She opened a drawer in the chest and rummaged amongst the cut-glass perfume bottles and eyeshadow, stuffing items into her Judith Leiber make-up bag. Slamming the drawer shut, she came towards me in a cloud of anger and Elizabeth Arden night cream, shoved me against the wall and slapped me.

"Let me tell you one thing, Edward. Up there on the farm, going about in your grandmother's old work clothes, that was the best time of my life. Our potato *lefse* with that strange brown cheese was the best thing I've ever eaten."

I crumbled, put my head in my hands. I cursed myself for not having told her everything. The weeks at Hirifjell had been real for me too. I went to stroke her hair. She shook me off, pulled out a cigarette. Ran her nail along the paper, pulled it slowly apart until the tobacco formed a pyramid in her hand. She opened the window and blew it all outside.

"Promise me one thing," she said, fixing me in the mirror. "That you never go into those woods."

I stared out of the window. Gwen kissed me on the forehead and said: "Go for a stroll, *mon chéri*. Go down to the Ancre and smoke a Gauloise. Do something dramatic. Summon the rain and the storm. And leave me alone."

The questioning look of the receptionist. The yellow autumnal leaves which flew about in the wind and pressed at the windows.

The bed was made, the sheets tight. On the desk were the photographs. An empty pack of Craven A. Over the back of the chair was Duncan Winterfinch's suit jacket.

She had left a letter.

> *Edward, I'm going home to Unst. Don't visit me. I've decided to put Haaf Gruney up for sale, to pay for the upkeep of Quercus Hall. Please spare us both any protest.*
>
> *Make sure you get home too. Keep Grandfather's jacket, as a memento. Of the summer when we were forever young.*

> *Gwen*

I sat on the bed, expecting an emotion which did not come. I wanted heartbreak, sudden, wild heartbreak. The urge to race to the train station as fast as I could, look desperately for her there, shout her name so it echoed along the platforms.

But it was like the longing for Mamma, when I wondered if I missed her as a son should. I witnessed my own sorrow for Gwen and asked myself if it was real. I sat alone in the hotel room and contemplated the emptiness; her suitcases had gone, her perfume still detectable in the air. But the urge was empty, the longing was false.

Perhaps because I knew it was not over.

She had left behind quite the gift, had Gwen: the question of why she had left.

Could the hidden photographs really have tipped her over the edge? Or had she used my secretiveness as an excuse to turn the tables, so I would not follow her?

I studied the pictures again, wondering if I had overlooked anything. Had something revealed to her where the walnut was? But there were no gloomy woods, no unfamiliar storage buildings.

Just a family on a road trip, at lay-bys, on motorways. And that one picture, covered with Gwen's fingerprints and handcream. Me and the toy dog in front of a brick wall.

What was I not seeing?

8

BEHIND THE STEAMED-UP GLASS OF THE GREENHOUSE I glimpsed movement. A figure amongst the plants, stopping every now and then, doing something I could not see. Only for a moment, when a drop of water gathered and ran down the glass, could I see her clearly.

Retired police officer Jocelyne Berlet.

As I approached the greenhouse I saw her blurry movements pause and she pushed open the door. She had the tall, slender build of a long-distance runner and made no effort to conceal her age. Streaks of grey hair were pulled into a simple bun at the back. The wrinkles around her eyes were deep and well defined, but otherwise she looked almost the same as in the newspaper photograph.

Jocelyne Berlet sized me up. The woman who had let me out of her arms and into life seemed to be asking herself if she could have done more for me.

"For many years I wondered whether I would recognise you," she said in French.

"And do you?" I said.

She studied me closely, not noticing that I had neglected to use the formal *vous*. Nodding briefly she said, "Yes. Your mouth and your nose. It's so strange to see you again."

I felt uneasy.

"How is your French?" she said. "Or should we speak English?"

"Preferably English. Even though I know some French from when I was little."

I bit my lip. Perhaps I had chosen the wrong words, said them in a way that betrayed my errand.

"Your French is still good," she said, and her English was more broken than my French. The humid air of the greenhouse wafted out. She kept long rows of vegetables and several roses inside. Air vents in the roof were regulated by an ingenious system of ropes and pulleys, and when she opened them a little, she did not seem completely satisfied and closed two again. I sensed similar minute adjustments taking place inside Jocelyne Berlet, a calibration of how much fresh air she could allow to enter an old story without its most delicate shoots wilting away.

"Let's go up to the flat," she said. She undid her apron and rinsed some rose shears in a water butt, pushed the steel into a bucket of sand and pulled them out again, shining and clean.

I looked into the bucket curiously. She switched back to French.

"Normal sand," she said. "With used motor oil mixed in. The sand scrapes off the dirt, the oil keeps the steel from rusting."

In the narrow dark entrance to her flat she hung my tweed jacket on a clothes hanger and I wondered if she noticed the label. If her years as a police officer allowed her to spot that the jacket had been owned by someone else.

"Let's sit in the kitchen," she said. "This is not a story for comfortable furniture. In fact, I don't own any comfortable furniture."

"What I know for sure," said Berlet, "is that your parents died at around six o'clock in the morning. The alarm was raised by a man who had been out carp fishing. He was on the far shore, a long way across, and there was thick shrubbery, but he was still a reliable witness. The best you could hope for."

I shook my head, perplexed.

"Because carp fishing involves patience and observation. You sit in complete silence. At around 6 a.m. he heard someone shouting, and saw a figure in a red jacket running towards the water. At first he thought they were other fishermen, someone jumping in to help reel in a big fish, but then it all went strangely quiet. He thought to himself: What do you do when you have a fish in your net? Either you take it ashore, or you try to catch another one. But the fishermen on the other side had disappeared. As if they had become invisible. The witness stood up and thought he could make out something red floating in the water."

The fisherman had then cycled home and called the police. I was finding it painful to listen to. She retold the story calmly, broke into French now and again, but always returned to English.

"And then . . ." I said in French, "you went into the woods?"

"Not immediately. According to what he told me, I found the place on the map and was rather surprised. It was a fenced-off wood, full of unexploded shells. So another officer and I took the inflatable police raft down to the fishing spot and rowed across."

"Were there no barricades?" I said. "A barbed-wire fence?"

"Here and there, but in poor condition. After the accident a new one was put up, much higher, but even today you would be amazed at how poorly secured the woods of the Somme are. We French are satisfied with a warning sign. But that's no help if you can't read it."

She scratched her arm.

"Who was wearing the red jacket?" I said after an uncomfortable pause.

"Your mother," Berlet said quickly. "It was a kind of anorak. Your father was dressed in dark blue. He was a little more difficult to find."

She stared at me, assessed me as Bestefar had assessed frayed rope.

"Eventually we got them back across on the raft. We thought they must have stepped on a shell and were surprised to see they had not been wounded. Not a scratch. Tracks in the grass led to the spot where they had been found, it was as though they had run into the water and forced themselves under, in order to drown."

"The gas," I mumbled. "*Le gaz.*"

"I'm sorry?"

"Gas shells. Wasn't it a gas shell . . .?"

Her index finger traced an invisible pattern on the table. "Yes, but we didn't discover that until much later, during the autopsy. There was a foreign substance in their lungs. Not mustard gas or phosphine, but a combination which the chemists had difficulty identifying."

"The fisherman didn't see anyone else?" I said.

She shook her head. "No. But the thicket was very dense. You've probably been wondering this your entire life, and it was the first thing I asked myself too: What were they doing in the woods? We weren't given permission to go in. We had to wait for the military, who arrived a few hours later. By then I had raised the alarm. Obviously we had spent the intervening time searching the surrounding area, and we found a car which we thought might be of interest to the investigation."

"A Mercedes?"

"Indeed. A black one. On a dirt track not far away. Norwegian number plates. The bonnet was cold, so either it had been there for some time or it had not driven far. I gave the order to break in."

Bestefar's words echoed inside me. *Someone wanted to get in once*, he had told me many years earlier when I asked why the Star had rusted scratches around the lock on the boot.

"What I found," said Berlet, "sent a shock through me. Toys and children's clothing. A small blue rain jacket. We found their names on some ferry tickets. Three passengers. We posted volunteers outside the woods, closed off the roads so the sound of passing cars wouldn't drown out the cries of a child. But because of the situation our primary theory was – I hope you can forgive me – that you too had drowned. While waiting for the sappers we took out more boats and searched for you on the surrounding land. And we kept wondering what possible errand they could have been on."

"Did you find our footprints?" I said.

"Yes, tracks from the car into the woods. I remember now that we partly contravened the order not to enter; we sent in a police Alsatian, which had your scent from your clothes. But the poor dog stepped onto the remains of a shell and had its hind legs blown off. It was put down immediately. We needed specially trained dogs, and until they arrived, there was absolutely nothing we could do. It was terrible, because we feared you too would step on a shell. The sappers wore gas masks when they went in."

"Did they find a toy dog in there?" I said, before realising how stupid it sounded.

"A toy dog?"

"Yes, a wooden one," I said, and thought about the poor Alsatian. "I think I had it with me, but it's gone."

She needed a few seconds to digest my surprising concern. "I imagine the search team would have found it," she said. "That is, if it didn't fall into the water. But only the carp can tell us that."

For someone living alone, Jocelyne Berlet had a remarkable number of photograph albums. They were stored on two shelves above her small television, visible from where we sat at the kitchen table. On the telephone she had told me that she stopped working for the police in 1975. What she had not told me in our brief conversation was that she had worked at an adoption agency after that. The reasons were so obvious neither of us needed to mention them. But I assumed I was in her very first album, one she stored in her memory.

Within me there was an album that was just as thick. The story of a master cabinetmaker and a one-armed timber merchant. I had a growing desire to tell her everything, but would she actually want to know? I did not want to be a key witness who arrived twenty years too late for a woman who had done the best she could. For her, peace had descended over the case, just as peace had descended for the adopted children in her albums. She followed their fates to a new home before turning away, for everyone's sake.

Because she never asked what else I knew. She told me what *she* knew, but with no renewed curiosity. Yet another person who let the past remain in the past. It has to be Einar's drive, his determination that lives within me, I thought, this urge to complete a race even though everyone around you is dead.

On the worktop there were fresh tulips in a glass vase, I recognised them from the top shelf of the greenhouse. She set them down between us. "I haven't even offered you something to drink," she said, opening a spartan kitchen cupboard. "Would you like some tea?"

She filled a scratched-up kettle and continued: "We were able to track their movements. They stayed at Hôtel de la Basilique and ate lunch at the Auberge. The restaurant was quite busy; a busload of American military cadets were on an expedition. A table had been reserved for three adults and one child. But the fourth guest never arrived. The staff thought they had seen a man approach the table, he seemed to be upset about something, but the waiter had been too busy to give a decent description. Then they ordered their food, asked to reserve a table for the following day too, and the fourth place setting was cleared away. That same evening the hotel receptionist saw them come in. It was late, and your mother was apparently terse and ill-tempered. You had fallen asleep, and your father carried you upstairs. Everything suggested that they were going to settle in for the night. But then at some point the night staff, half-asleep, put a call through to your room."

"Someone called in the middle of the night?"

"Yes. This opened up a number of possibilities. Such as them agreeing to meet someone in the woods."

"Did you find out who phoned?"

"No, only that it was a man speaking French. It was a strange time to call. We even asked ourselves whether it could be someone from Quebec who had forgotten about the time difference, but we thought that unlikely. The night clerk hung up, but then a guest from a neighbouring room called to complain about a child crying. The clerk got dressed and went along the corridor. Everything was quiet. So either you had gone out in the interim, or you had fallen asleep again."

I pulled myself together and said: "Which room were we in? I'm staying at the same hotel."

Jocelyne Berlet studied me for a long time before slowly shaking her head. "I don't remember. Unfortunately. But there are maybe only ten rooms there, so the chances are . . . well, in any case, we found their clothes and luggage inside. There was no indication of a hurried departure. The toothbrushes were in the glasses, the beds were tidy, the clothes neatly arranged. The room had been booked for four days, and the staff wondered why the three of you didn't come down to breakfast."

Back home at Hirifjell I had my preferred grounding points. The conservatory, where I would sit facing the spruce trees at the edge of the woods and the lone pine with the magpie nest. A good place to be when I needed to think. In the classroom it was a wall chart, one that was never rolled down but hung between the world map and the map of Norway, yellowed from the sun streaming through the windows during the summer holidays.

Jocelyne Berlet had a similar point in the form of a junction in the water pipes on the kitchen wall. It was close to the floor, just above the skirting board. I got the sense that she was used to sitting here like this, her hands folded around her left knee, while her reasoning flowed freely. She thought for a long time before loosening her grip on her knee and her eyes wandered back from the pipes.

"Let me ask you one thing, Edouard. Your reason for coming here. Are you here to find peace? Or . . . are you searching for something?"

"This probably sounds strange. But I have to know what happened."

"Yes, but *why*? Again, excuse me. *Why* do you want to know this? There's a reason we prefer to remember the good things that happen in our lives. People have a phenomenal ability to filter out the bad. So the bitterness is given a sweet dose of healthy reality. Children can adapt to most things. But now you're returning to the place where your truth once existed. That could open wounds you may not be able to bear."

She was right, of course. She had said that Mamma was upset when we returned. While I had told myself that a sad person does not eat perch with saffron. So maybe my only true memories were those I believed to be true.

"Are you hiding anything from me?" I said.

"Not at all. But there are theories about what happened, which may be too much for you."

The water boiled and she switched off the hotplate.

"It's so . . . unfair," I said. "I know of people who went into those woods in 1943 without getting hurt. Yet Mamma and Pappa stepped right on top of a shell."

She shook her head. "But that can be explained. The shell casings are made of iron. They rust, but it happens slowly. During the previous war they might have been stepped on without consequences. But by the time your parents went into the woods, the shells had lain there for more than fifty years."

"And now?" I said.

"Now they are even more rusted. It's nature at work. The woods are getting more and more dangerous."

She could tell what I was planning.

"Don't even think about it," she said. "People still die from unexploded shells. We had an incident near Thiepval only a couple of weeks ago. Two road workers were taken to the hospital, confused and groggy. They had somehow displaced a gas shell that was so rusted there was only a couple of millimetres of metal left. In fact it made me think about the incident – your incident – when I read about it."

She stopped to glance at me out of the corner of her eye. An unsolved mystery that had grown up, a young boy who was now a man, sitting at her kitchen table in the flesh.

"What do you think actually happened?" I forced myself to say. "If you could imagine a scenario."

"I have a theory. But to understand it, you'll have to understand the area we live in. The Somme releases the *loss* in people. The battlefields leave such a violent impression that it puts some people in quite a state. The history of this place creates a longing for meaning and life, for something humane. I know what it's like not to be able to have children. The empty space inside and outside of you. So I believe you were kidnapped, at the site of the worst slaughter in the history of mankind."

"You think that someone just . . . took me?"

"Time after time I have seen this desperation at the adoption agency. Even in its Sunday best, with a forced politeness that attempts to conceal years of longing. So yes, I believe that you were separated from your parents, and that they went looking for you. I have no idea how it happened. Presumably you just wandered off and someone found you, scared and desperate. Maybe the initial thought was to get you to safety, but then another desire was awakened. There are millions of people out there who lack foresight. Something might seem like a good idea for a few minutes, but then comes the reflection. They ask themselves what the neighbours will say. Then the child gets hungry and starts crying. They think about the birth certificate they don't have. Their enthusiasm for the plan soon fades. Most often they come to their senses within half an hour, or a couple of days at most. Then there are those who think there is money to be had from missing children. But a plan like that would have gone to pieces after twenty-four hours, when they read in a newspaper that your parents were dead."

"The hotels would probably have asked why someone was suddenly showing up with an extra child?" I said.

"Exactly. You were gone for several days and whoever took you must have stayed somewhere. That's why we checked with all the hotels and guest houses in the area. We suspected an abduction by a childless couple in their thirties or forties. Or a single woman of the same age. We checked all the guestbooks in the district. Apart from exposing nine instances of infidelity, we found nothing. I think they must have driven you to Le Crotoy because it was far enough away from the centre of the search. They waited until Monday morning because all the offices were closed at the weekend. They decided on a doctor's surgery, a place with intelligent, responsible people. They might have sat a distance away and waited for the police car to arrive. They're probably still alive. With or without a child. What I am quite certain of is that they live far, far away from Le Crotoy."

"Do you remember the name of the doctor?" I said.

"Unfortunately not. Just that he was old and distinguished. Someone had knocked on his receptionist's door, and when she came out you were alone in the waiting room. They examined you at once. In fact they recognised you from the newspaper. Discovered that you had been fed recently and were calm. Your clothes were dirty, but your only injuries were the bruises."

I frowned. "Bruises?"

"Yes. Lots of them."

"Where?"

Jocelyne Berlet stood up and walked around the table. Stopped in front of me, and I saw her straining to keep her emotions in check.

"We withheld some details from the newspapers," she said and touched my right arm. "They were here." Her finger ran down my shoulder and upper arm. "Heavy bruising. The skin was almost black. But nothing on the lower arm, nor on the left arm. Violence must have been used."

Her hand rested there for a second too long. I was going to place mine over it, but she took a step back.

"I sat with you," she said. "I tried to get you to speak. To say something that could help me pick up the trail. But you were silent. It was as if the incident had blocked off your memory."

That was the day I stopped speaking French, I thought.

"And then . . . my grandfather Sverre arrived?"

"I saw the two of you reunited. You ran towards him and he took you in his arms. He had to stay in Amiens for a few days, to take care of the formalities. He had to identify your parents and their bodies were sent to Oslo by plane. Your grandfather was rather unsentimental. But he seemed to be somehow ... expectant, as if he was waiting for something else to happen. Then you both drove off in the black car."

I pictured him. The hate forged that day for the brother who wanted to do good, but who made a mess of everyone's lives around him. Like a dog wagging its tail too eagerly and smashing the porcelain to pieces.

"I seem to have forgotten the tea," she said.

Out on the street I lit a Gauloise and looked at the Bristol. I was not sure whether to tell her about Oscar Ribaut, but in any case the occasion had not presented itself. Her work probably trained her to be good at saying goodbyes. She nodded briefly when we stood by the door, neither of us allowing any feelings in or out.

I stood smoking with my back to her flat. A couple of buses passed, I thought I heard someone shouting, and then I felt something hard hit my shoulder, followed by the sound of an object rolling on asphalt. Jocelyne Berlet had thrown a ceramic fuse from the second floor, and hit her target. She stood at the kitchen window, her hands on the sill as if holding on to the railing of a ship.

I came back into her hallway.

"You were standing there smoking," she said. "Is there maybe something you haven't told me?"

"Maybe. You too?" I said.

She picked up a black leather glove that had fallen to the floor. "The explanation we leaned towards was of a spontaneous kidnapping," she said, putting the glove next to another on a small table. "That was the only possibility that seemed to settle matters."

"But did it?" I said.

"No. For me, it was never settled."

The prospect of a development hung in the air. Some exchange of information to resolve an old case. So I took out the photographs and told her everything I knew. She studied the pictures of me and the toy dog with an expression similar to Gwen's – sad, but also affectionate. A young boy a few days before he disappeared, before his memories were erased.

Afterwards, Jocelyne Berlet sat with her fingers rubbing her temples, as though willing her brain to pick up things where it had left off. "This has not given me any new insight," she said. "But there are irregularities in any investigation. For example, they had two sets of keys to the room but we only found one. In your mother's jacket pocket."

"I think he was supposed to look after me that day," I said. "The man called Einar."

"Either that," she said, "or they lost one set of keys while they were looking for you. It's impossible to tie up all the loose ends. The other thing that bothered me was the dress."

I froze. Felt my neck tightening. "The dress?" I mumbled.

"At one point we suspected a cleaner of pilfering from the hotel, but we let it go. When they were found, your mother and father were dressed in normal travelling clothes. But the previous evening the waitress at the restaurant had noticed your mother wearing a beautiful, blue dress in an old-fashioned style. Your father – what is it? Is something wrong?"

"I . . . no, go on."

"Your father wore a suit. When we searched the room, we found the suit but not the dress. I was surprised. Who, shortly after someone has died, would enter a dead person's hotel room to steal an old-fashioned summer dress?"

AN ELEGANT COUPLE WALKING THEIR DOG. I WISHED I
had the Leica now, to make me just another war tourist.

"I'm trying to find a farm," I said, "that once belonged to the
Daireaux family."

They looked at one another, shook their heads. Perhaps I
had pronounced it wrong. I tried again, rounding the As. They
shrugged and continued on their way.

I had left the Bristol on the side of the road. The dull headlamps reminded me of the gaze of a faithful old dog. Slow to
turn, but with a nice, comforting smell. If only it could tell me
what it had experienced during all the miles it had carried Einar.

I swung open the door and sat with my legs outside the car.
Found my cassette of Bob Dylan and fast-forwarded to "Mr
Tambourine Man". I bent to tie my shoes. I had been sloppy that
morning, only a single bow, but now I folded the laces for a Turquoise turtle knot. *That* was why Gwen could make a knot that
held the entire day – because a one-armed man cannot tie his
own laces.

Somehow sensing that she was still close by I crossed a small
bridge over the Ancre and walked uphill as far as a disused railway track. On the other side of it was an old brick building with
an arched roof. In the picture of me with the toy dog there had
been a brick wall in the background. But how many hundreds of
buildings like that were there?

I walked around it, kicking at the metal junk buried in the
grass, pulled myself up to the windows and peered inside an old

waiting room. A massive clock on the floor, smashed. Nobody to be seen, only the sound of birds and traffic in the distance.

I pressed on and came across a stooped older lady standing in a ditch with a yellow beach ball. A young girl came running out from behind a hedge. The old lady gave her the ball and laughed.

She had been alive long enough to know the local history, and she pointed me in the right direction. "But the Daireaux family — they are gone," she said.

"Did you know them?" I asked.

She shook her head. "Others lived there after the war, but they must have moved a few years ago."

I followed her directions. If she was right, the farm was a couple of kilometres away from the walnut woods.

I could just make out an old dirt road between some gnarled trees. It was barely passable; the spring thaw had used it as a stream, and tall grass had grown up in the middle.

I had a fleeting image of Mamma and me walking together, a phantasm that came from here, and mattered here; because the two of us had wandered this path separately, but with similar thoughts: Does a part of me belong here? Or could I have grown up anywhere?

Suddenly the vision was gone, and I was clinging to something yet still losing my grip. Like a pine cone that only realises it was connected to a tree when it falls and scatters its seeds.

Fifty metres ahead I saw the Daireaux farm. A farmhouse with a sunken roof. A barn with no gates. An outbuilding with the doors hanging crookedly from their hinges. The descriptions from Einar's letters fell into place. The henhouse. The stairs leading to the main entrance, where my great-grandfather met the mysterious Oscar Ribaut.

I sat on the top step and surveyed the courtyard. Here they had stood, Einar and Isabelle, hoping that the war would end. And the Gestapo had come here, right here, and arrested her and her family while Einar escaped.

The roots of the bushes had grown into the walls and the foliage had stained the paintwork. A tarpaulin flapped in the wind. In the barn there was a dead pigeon, the floor of the stables was covered with greyed straw and mouse droppings. On the kitchen worktop in the farmhouse there was a bicycle wheel. The smell of food had not filled that room in many years.

I wanted to be someone the dead could rely on, but none of the dead had come forward to help me. From time to time I was forced to ask myself whether, if I found the walnut and got my hands on the money, I would buy back this farm. Would I use it as a summer house, squeeze the last drop of blood out of the tragedy and show the family ghosts that the circle was complete? Now that I was here I wandered around in the hope of recognising something, a belonging, a responsibility. But nothing touched me. It was like looking at a picture frame from which the painting was gone.

Only when I left the Daireaux farm did I feel a sadness.

That's something at least, I thought.

On the way back to the car I felt as though I had all the time in the world. The locals did not stare, they seemed to be used to all the tourist buses, to people trotting about with cameras pointing across the landscape, to English cars stopping in the middle of the road.

At a small graveyard, some schoolchildren were jumping between the graves and shouting. I wondered what Duncan Winterfinch would have made of it. At that moment a taxi raced past, slowing near a side road on the other side of the graveyard. The passenger leaned forward to speak to the driver.

It was Gwen.

Wearing a new, dark-blue jacket. With her hair styled. She was fumbling with a map, pointing at something. The car drove on. I set off in the Bristol, but I did not find the taxi, and instead went into a cafeteria with white plastic furniture and harsh lighting. With a lump in my throat I remembered her face across the table at the Raba, the smell of her skin and the way she would sigh before she fell asleep, the radiance and willfulness when she planed *Zetland*. Now it was over, empty and dead, gone like the trees on Shetland.

An hour later I was standing by the Daireaux woods. Using the war map, I had followed a dirt track through some rocky fields and stopped where I imagined Mamma and Pappa must have parked the Mercedes. But I did not stay long, I would be back the next morning.

I drove on, looking for Duncan Winterfinch's summer house. Another round of the cemeteries, through creaking iron gates and amongst gravestones, taking in the tragedy of 1916 and trying to put myself in the head of the old captain. On a stone wall there was a brass hatch I had not noticed earlier, with a cross embedded in the metal. I opened it and found a slim book inside the niche.

A visitors' book, with names, countries, dates. And comments, occasionally. I wrote my name on a new line. Carefully, so the pen would not cut through the paper. *Edouard Daireaux*. I replaced the book and studied the gravestone of an unknown soldier. *Known unto God*.

A thought struck me. I returned to the brass hatch and took out the visitors' book again.

The local office of the Commonwealth War Graves Commission resembled an upmarket tool hire company. There were lawn-mowers, mini diggers and cement mixers in large, well-ordered sheds. Only the allied flags, swaying on tall poles along the drive, alluded to a sense of common history. A large truck had turned in and was now idling as a man in overalls hopped out of the passenger side and picked up some reels of electric fencing. An animal transport vehicle, judging by the ventilation slots at the sides and the plaintive baaing.

Sheep made up the majority of the employees here, I had seen flocks of them by the memorials. They grazed the old trenches, which were too dangerous to reach with lawnmowers.

The sheep transport drove on. I dispelled the thought of the animals at home and knocked at the office door. No answer, but in I went anyway. At a table further inside sat three men and two women in green work clothes. On a flip chart was their work schedule. Their job was to keep the war memorials tidy, maintain the plants and change flags that were bleached by the sun.

"The visitors' books for the war cemeteries," I said in French. "Where are they kept when they're full?"

They looked at me sceptically, and then one of the men gestured for me to follow him to a basement, where he unlocked a steel door. A fluorescent tube blinked on above us and others lit up inside the endlessly long archive room. A handwritten collection of the many forms Loss has taken over the past seventy years. Worn and weather-beaten visitors' books like newspapers left in the rain, their pages buckled.

"630,000 fell here," he said. "That's only counting the allies who fought in this battle here, in 1916. The Germans have their own burial grounds with just as many."

He asked why I wanted to see them, thousands of record books with pages that had never been read, but which were too precious to be thrown away. I told him that my parents had died here in 1971, and that I was searching for their last signature.

That was only partly true; I was also looking for another signature, of a one-armed man. The employee showed me the shelves with books for the cemeteries around Authuille, around High Wood, all the places where the Black Watch fought in 1916. Hundreds of volumes for each cemetery, with names densely written in fountain pen, pages torn by the wind. The signatures from the twenties were neat and formal, sometimes hundreds in a day. People must have queued. One-line messages: *We miss you dearly. Sarah is nine years old now and is doing fine.*

Parents of soldiers. Widows of soldiers. The rightful claimants of an empty fund in Scottish Widows.

I pulled out a record book from the thirties, then another from 1953, and noticed the developments in how British schools taught handwriting. I read how the pattern of grief changed as parents grew older or passed away; how the legacy of war which began as a bloody family tragedy, turned into a historic, national event.

But I could not read for long. I skimmed the pages in search of a majestic signature I had seen before, on the contract giving Einar the right to reside on Haaf Gruney *in perpetuity*. Perhaps the signature he regretted most in his life.

But the columns were endless, it would take hours to find him. I skipped ahead to September, the month in which the battle of the Daireaux woods was ultimately won.

And there he was. *Capt. Winterfinch. The Black Watch.* A signature with rank and regiment still, despite the fact that he ran an international business with offices around the world.

The visit was repeated year after year, and always in September, which must be the wettest season here; the autumn record books were spotted with mould, with curled pages and signatures marred by pens which would not write. From 1928, the timings of his visits became more regular. He was one of the earliest to arrive in the morning, perhaps to be alone with his grief. His signature filled two lines, and he was never accompanied by anyone.

Each year Captain Duncan Winterfinch had visited his fellow soldiers' graves, and from 1931 onwards he was invariably the very first visitor. First he had driven to Thiepval, then methodically to each and every cemetery, along the endless white gravestones. To be on the safe side, he always brought his own pen, a wide-nibbed fountain pen with green ink.

I turned the pages more rapidly, past the five missing years of the Second World War, and found his name again in September 1945. The signature was a little more slanted that year. Had he passed the Daireaux woods, seen the hollows where the trees should have been? It must have felt like a combination of betrayal and blasphemy.

1968, 1969. Still he was the first, still he signed as Capt. Winterfinch. 1970. I knew his system now, and opened the record book for the autumn of 1971, to October. Flipped backwards. I was approaching the beginning of my existence, my own year zero.

There I discovered that the last time Duncan Winterfinch had visited his fallen comrades was early on the morning of September 23, 1971.

This time, for the first time, he had brought someone with him to Thiepval. Beneath his signature was one that I had not expected to see, much less her comment. Because there it was, Mamma's last testament, written just before she died.

N. Daireaux. May you find peace.

10

I WOKE AT THREE IN THE MORNING AND FELT A PRES-
ence in the night. Switched on the light and stared at the white
boarded walls, then at the telephone, half expecting someone to
call. Left myself open to the possibility that the ghosts of twenty
years ago had visited me.

But I was the only one here. Only I knew everything that had
happened that morning, only I had heard every word spoken.

I got dressed, put on the kettle and made a strong cup of instant
coffee. I would never be able to re-create that day's events exactly,
but I knew *enough*. What I needed was a truth that I could be content with.

They must have arranged to meet Duncan Winterfinch at the
restaurant, but it was too busy, full of people talking at the tops
of their voices. A man who built a palace of oak in the most desolate spot in Great Britain would not be prepared to negotiate
in such a clamour. It was the most important transaction of his
life, when the wood from the sixteen trees of the Somme would
be brought to light again. He left the restaurant exasperated and
indignant, and that night decades of pent-up anger must have
gnawed at him. According to Gwen he had always woken at
three.

Duncan had called the hotel room that night. A loud ringing
from an old telephone. He would have suggested that we meet
at Thiepval, at the crack of dawn. There, free from the droning of
cars, free from curious eyes, with no noise other than the twitter-
ing of birds, he would describe the advance of the Black Watch in

1916. A war veteran's final story. The same temperature, the same air, the same light.

The British believed that a soldier owned the ground on which he died. This was something Einar could never understand; he was blinded by his and Isabelle's story from those same woods.

He may have been amenable, Duncan Winterfinch. His sleep was out of sync and Mamma may have suggested it. Let's meet *now*, she might have said. If a three-year-old boy is wide awake anyway, there's no point in sitting in a hotel room waiting.

Outside it was a chilly morning. I started up the Bristol and drove through the dawn. The streets were empty, the gilded church spire barely visible against the sky, just as it had been twenty years ago. A wide-awake child, sleepy parents, on their way to a war memorial to meet a one-armed man. I tried to evoke the memories of the street lamps reflected in the bonnet, the humming of the engine, the yellowish gleam of the instruments. Mamma and Pappa on their way to die.

Up on a hillside I could make out the silhouette of Thiepval. A colossus of a memorial, impossible to determine if it was ugly or attractive, built for eternity to broadcast the cries of the dead.

The car park was empty, the only sound my footsteps on the gravel. The monument rose in the dark as I walked towards it. Think. You *have* been here before. Remember the shape of the arch against the sky, the birdsong in the distance?

I stopped to breathe in the raw morning air, and it occurred to me that Gwen might be here too. I reached the wide steps of the brick colossus and climbed amidst the cold aroma of consecrated stone; I was there with the dead. The sound of my steps smacked hollowly beneath the arches. Everything was dark, chilly, old.

Ten minutes passed as dawn crept forward, the morning light glimmering on the stone surfaces, summoning a sea of engraved characters, row upon row. Then the sun came up, and seventy-three thousand names emerged and surrounded me.

I sought out the stone panel for the Black Watch. They had

stood here, Winterfinch and Mamma. He had the company behind him, and she had a history that began in Ravensbrück. They had plenty of time for long explanations at this hour of the day – for a man who wanted to get his house in order before he died and a woman who bore responsibility for a mother she did not remember.

I did as they had done. At the stone steps at the bottom I found the brass box built into the wall with a cross on the outside. I took out the visitors' book and signed it, the first person that morning.

Why had they not settled it there and then? Why did they have to go to the woods? Perhaps he made Mamma an offer, suggested a sum of money. Perhaps they had gone for a drive in the car, to be alone, to have time to think.

I left Thiepval and drove to my family's walnut grove. It was about the same size as the flame-birch woods, but surrounded by scrub and undergrowth which pressed against the fence. There were some beech trees growing a little further in, the crowns visible against the bluish sky. The birds were silent. There was the scent of damp earth in the air.

By the rusted barbed wire I felt suddenly unwell and was almost forced to my knees. My clothes clung to the cold sweat on my back. I pictured Mamma and Pappa's names on a distant, weathered grave, carved in blue Saksum granite, a grave I could not bring myself to leave flowers on.

Then it was as though the woods answered me. The wind rushed down, carrying autumn with it. A whistling from the trees. The flutter of a precious memory.

Pappa's hand holding mine. Firm and strong. In the other, something new and amusing. The wooden dog. Now I carved out another memory, no detail, just a feeling of disquiet: my impatience. Because someone was talking and talking incessantly. Then the three-year-old boy thought: It's about time I got some attention. The restless child who hides behind the apple trees every morning. *And no matter how tired I am, I join in, because every time it is as if we rediscover each other, it is a reminder that life has meaning now.*

We were never meant to enter the woods. The agreement had been made, they had come here to cross the final T. Pappa's grip loosened for a brief moment, and then we lost each other for ever. Suddenly I was off and running. I ran like I had my entire life, in through the tree trunks.

The truth was that I had killed my parents. If I had not bolted into the woods, they would still be alive today.

I forced myself to enter. My heart wanted to leap out of my chest, and my body was so reluctant that my testicles contracted. But I was only one of tens of thousands who had felt exactly the same way in the woods.

I climbed the fence and lowered myself slowly down the other side. I kept so close to the fence that my clothes snagged on the wire. The branches reached out and brushed me with dew. The path was impossible to find. It was likely that no-one had been here since 1971. I continued along the fence until I reached a belt of old aspens.

Aspen. The first tree to grow in a clearing. Or on a path which is no longer used. I left the fence, bending twigs back and allowing the line of aspens to lead me in past overgrown trenches, shell craters and snapped trunks, through a wilderness in which nature was trying to cover its war wounds.

I came to a clearing. A bare, dead stretch where grass grew reluctantly. Sixteen large hollows, narrower and with straighter sides than the craters left by the shells. The roots of the walnut trees. Thousands of dead soldiers beneath me, kneaded into the earth.

All was still. I realised that even if I had strayed here without knowing the history, I would have felt exactly the same: this was no longer a forest, it was a mass grave.

Here and there around me were what appeared to be small piles of stones. They were overgrown with decayed foliage, but I glimpsed rusted metal in between. The heaps of shell casings the sappers had collected.

Beyond them was a crooked tree, with large nuts on the ground beneath it. It must have germinated before 1944, a descendant of the Daireaux family's walnut trees.

The nuts appeared fresh, like those we ate at Christmas, the shape and veined ridges resembling a brain. I put some in my pocket. Further ahead through the bushes I could see the glint of water.

Twenty years ago I must have run towards the water. In the night I had convinced myself that I needed to see where it happened, but now the tendrils of my common sense grabbed hold of my legs and I was reminded of the Alsatian with the severed hindquarters. The crackle of dead foliage. I heard a voice telling me to go back to my very first memory. To remember that more than anything.

The journey led me to Hirifjell. It must have been the first winter without Mamma and Pappa. I sat on the snow crust looking at Alma and Bestefar who were standing by the front door, lightly dressed under the porch lamp. They said something, and I realised that it was about me. But what seemed important was that the crust was hard and could support my weight. And that spring lay beneath it.

I gazed through the trunks at the glittering water at the bottom of the valley, beyond a no-man's-land of gas shells and certain death. And then I understood what I had to do. It was as if I had walked with a loose wire in my hands all my life – now I realised that it was a bowstring, twisted from an umbilical cord.

I fought my way out of the woods and ran through muddy thickets to the other side of the pond, until I came to the place where the fisherman must have stood. Trampled ground, discarded fish guts and cigarette butts revealed that it still was a good spot for fishing.

I threw off the tweed jacket and removed my shoes and trousers, then stepped into the water. My bare feet sank straight away, the stink of decay bubbling up. A toad hopped off. I stretched out flat in the murky water, the algae on the surface covering me to the shoulders as I kicked my way through the mire.

It was a couple of hundred metres across. All I could hear was the splash of my strokes and I turned to see my zigzagging path through the algae. I had the feeling I was being watched.

I approached the far shore where the water was shallower and flowed in a swifter stream. The trees hung over the water, their branches breaking its surface.

This was the place in which they had died.

And the water carried me. I floated above the shells as I combed through the past, each stroke drawing open infinite veils in my mind, until finally I came to the last.

I remembered.

I remembered.

Not one coherent event, but a series of deep impressions.

Mamma and Pappa there in the thicket. Two figures shouting my name, struggling to get through, and at the same time, inside me, then as now, a feeling which went from excitement to horror and darkness.

They bounded towards the shore, and then a poisonous green substance rose up. Mamma doubled over and tumbled down, tried to grab hold of Pappa, they got up together but fell again, onto the muddy shore and into the water. Desperately they tried to scramble back onto the shore, but their bodies would not obey.

It took them a long time to die. Pappa was the last to give up the struggle. The surface of the water grew calm. Pappa lay with his arms spread, the water rippling above his forehead, a faint vibration tracing the thoughts inside. Mamma was on her side, with her hair outstretched, pointing in the direction of the current. Her face was peaceful, her eyes were fixed on the shore, and they said:

You are safe, I can die now.

After that, nothing. Just a great empty void.

In a trance I swam back, grabbed an overhanging branch and

pulled myself up amongst the fish guts and flattened cigarette packs.

My heart was pounding. I tried to dry myself with my shirt, but realised that I was sweating out of fear, I could not get dry. Then, putting on the old tweed jacket, it was like a flash shed light on a brief but vital memory. A particular smell, a combination of nervous sweat, damp tweed and Balkan Sobranie.

I remembered no more. But it was obvious what had happened.

Mamma and Pappa's shouting. A muffled explosion in the woods. A strange figure who approached and took me away.

After that a barrier rolled away in my memory.

Duncan Winterfinch had saved me. He must have followed Mamma and Pappa when I took off, perhaps helped them search for me. Then heard the whistling of poisonous gas in the distance and had seen the green fog that surrounded Mamma and Pappa, recognised their confusion. It had happened again. As meaningless as what he had experienced day after day in 1916. People dying. Rescue impossible.

Then I must have popped up from my hiding place. Save those who can be saved. He lifted me up with his one arm and pressed me to his chest so hard that I was bruised. I had replaced his severed arm. My nose against his smoky jacket.

Then we ran across Speyside Avenue. There was no sheltering from the poisonous gas, we had to get away. He threw me inside the car and drove off. A frightened old man, the head of a timber empire, with a crying child – where could he go at six o'clock in the morning, and he blaming himself for the deaths of two people? To his summer house, most likely.

Later, Einar must have come to Authuille. He would have heard the sirens, discovered that there were two dead and a child missing. He had let himself into the hotel. Stood in the empty room and allowed the certainty to filter in. Did as he had done in 1944. Took the only precious keepsake with him, with the scent of someone dear to his heart: Isabelle's summer dress.

Then he had gone to the one person who might know something: Duncan Winterfinch. Maybe I ran to him, happy to see a familiar face. A rare occurrence for Einar, someone eagerly running towards him.

It was then that I remembered something from Einar's letter. In 1944 he had gone into hiding with a friend by the name of Charles Bonsergent. *He came from a fishing village a day's journey away.*

A family of fishermen. And where did they live, people who subsist on fish generation after generation? Not near fields, but in a place like Le Crotoy.

11

A DAY'S TRAVEL IN 1944, BUT ONLY A COUPLE OF HOURS in a car.

Le Crotoy reminded me of Lerwick. The smell of the sea, the boats setting out. I had not expected to discover much, just a hint of evidence that this was indeed where Einar had taken me. I had driven along the bay of the Somme, a gigantic wetland with sand dunes, bogs and stagnant water. Duck-hunting terrain where you could wander for weeks.

A narrow main street, some houses near the sandy coast, small shops and a closed school. Trudging around looking for a doctor's surgery, I passed two fishermen on a bench. I was about to ask directions but their expressions were closed. I walked on to the kiosk at the viewing point by the sea; it looked like it had been there for at least twenty years, as did the plump lady who ran it.

"Gauloises," I said. "And a lighter, please."

Without turning she reached behind her, picked up the correct pack of nineteen, and placed it on the counter.

"How many doctor's surgeries are there here?" I said.

"Two," she said. "Are you trying to give up smoking?"

"Not for the time being, no. Actually there is one old doctor in particular I'm looking for. Someone who was here in 1971."

"In 1971? We only had one then. Docteur Boussat. He worked here until the day he died."

"When was that?"

"In 1980, I think, or 1979. I don't remember. Maybe even '78."

"Tell me, where was his surgery?"

"It's closed down now."

"But where was it?"

"Next to the Citroën garage," she said and pointed through the hatch of the kiosk. "I went there a lot when my children were little, they both came down with false croup. Just follow the street until the end."

The waiting room in which I had been found was now the garage hands' break room. At first the four mechanics sitting there told me I was in the wrong place, and then they could not – or would not – show any interest in my errand.

They wore safety boots and oil-stained boiler suits. Free men who did not need to shave or answer to anyone; their fellowship harboured suspicion of an outsider like me.

"What do you want with the doctor?" one said. "He died long since."

"He helped me when I was little," I said. "I just wanted to see his office again."

The men shook their heads. What had happened in the past was of little concern to them.

Bare walls, stools around a Formica table. The Pirelli calendar left open at a summer month. Somewhere near here I had been abandoned. Now there were only overflowing ashtrays, spare parts in the corners and oily floors. I could have entered any room and felt the same. Nothing.

They brushed the ash from their boiler suits and stood up, all but one, an older man with a crooked mouth who sat at the far end of the table. He had not said a word, either to me or to the others, but now he began to talk with a stutter, presumably fearful of being sneered at.

"Doctor Boussat helped me after the war," he said slowly. "The Gestapo beat me with a baton. I used to be a fishmonger." He forced out the words. "But as a mechanic, I don't need to talk very much. Why are you looking for Monsieur Boussat?"

There was no harm in telling him. And then we looked around the room, as though it had once again become a waiting-room – for him too.

"I seem to remember reading something about a disappearance in the newspaper," he said. "But I doubt you'll find anything here now."

It was then that I noticed in the corner a dark-brown chest of drawers in a familiar *art-deco* style. It had been spared the scratches of engine parts and tools. On top of it was a collection of old Citroën spare-part catalogues. I crouched to look underneath it, but it did not bear Einar's distinctive signature. Instead there was a small fish, the initials c.b. and the year 1940.

Charles Bonsergent. And this town was smaller than Saksum. The fewer streets a town had, the longer a reputation could survive.

"It was left here after Boussat passed," the mechanic struggled to say. "We didn't have the heart to throw it out."

"It says c.b. here. Could the cabinetmaker have been Charles Bonsergent?"

At first he did not understand, but then he looked at me for an explanation as to how I knew the name.

"Did you know him?" I said quickly.

"No, I wouldn't say that." He collected himself for a few moments. "But of course I knew who he was. Le Crotoy is not a big town."

The mechanic had work to do, but he seemed to be curious as to why I was here.

"Charles was several years older than me. He came from a family of fishermen, but he was a skilled craftsman. He went to Paris and became a cabinetmaker. He was there when the war broke out. Back here, he made furniture for the locals."

"Was it he who made the furnishings for Docteur Boussat?"

"Quite possibly. I remember they were attractive – everything in the same style as that chest of drawers."

He was waiting for me to tell him something more, so that our incomplete stories could correspond. I asked if Charles had a friend from Norway. Or one called Oscar Ribaut. The mechanic shook his head, he did not know him that well. I could tell that his curiosity was beginning to wane.

"Charles was only a cabinetmaker in his youth," he said. "He made nothing else after the war. He was in the resistance, was arrested by the Gestapo. They tortured him, cut off two of his fingers. Something was destroyed inside him, he developed a tremor. Like my lip, but it was in his arms. Not that he gave up. They had a big house where they hid saboteurs right up to the liberation. But that was the end of his fine furniture."

"Interesting," I said, knowing that I now had the truth and would have to make peace with it.

I tried to put myself in Einar's situation. With a child. Dazed by the tragedy. Nobody around him who would understand. Everything turned on its head. Nothing that could be undone. His daughter was dead. His nephew was dead. His grandchild was terrified. Nothing for it but to go to a friend, someone who had helped him before.

It was like a continuation of his escape in 1944, once again out here to Le Crotoy. Perhaps Einar had rung Bestefar himself to relay the message he knew would bring his brother to his knees, and forge a hatred that would last his entire life. He would also tell him that I was safe, that Bestefar had to come here, but not until Monday morning when the doctor's surgery opened. Feign ignorance so as not to weaken his case as guardian, and not reveal that the kidnapper was his own brother.

Charles Bonsergent made a deal with the doctor, and in doing so ensured that Einar had some time with me. As much time as possible before he had to return to the rock and rain of Haaf Gruney.

The four days were no longer the extent of a mystery; they were a measure of Bestefar's sense of fairness.

The mechanic sat looking at me, and now he broke his silence. "About Charles," he said. "After the war Docteur Boussat tried to help him, but he was never rid of the shakes. So Charles took over his father's business and worked as a fisherman."

"Is any of his family still alive?" I said.

"Docteur Boussat's or Charles'?"

"Charles'."

"They must be."

"But do they live here?"

He shook his head. "They went away. Charles was the last of the fishermen. Odd story, in fact."

I had slipped into a reverie. The mechanic got up and motioned that he had to get back to work.

"How so?" I said quickly. "An odd story, you said."

"Well, the first thing the Germans did when they got as far as the coast was to destroy all the fishing boats. Burned or crushed them, so that people couldn't escape to England or smuggle weapons. After the liberation Charles was the first to start fishing again."

"But was that so odd?"

"No, it was more the boat. He built a large rowing boat so he could cast out nets near the coast. A boatbuilder showed up to help him. I have no idea where they got hold of the materials so soon after the war. There was hardly a plank of wood even for the crosses at the cemetery. But the boat was a popular sight. People were hungry."

The mechanic had a pen in the pocket of his boiler suit, and I asked if I could borrow it. I tore open the pack of Gauloises, emptied the cigarettes into my shirt pocket and drew *Patna* on the reverse side.

"A boat like this?" I said.

He looked at my sketch with his head cocked. "That's just a perfectly normal boat. Any old boat around here looks like that."

I crumpled the paper into a small ball and stood up to look out of the window.

"His first boat," I said, "is it still here?"

"No, he only used it for a while. Sorted out a larger one the following year, a more seaworthy vessel. Just as well, his wife didn't like the first one. He moored it on the sand banks of the Somme, just where it flows into the sea. An inaccessible stretch, filled with driftwood. I went duck-hunting there in the seventies. It was gone by then."

I asked why his wife had not liked the first boat.

"There's a tradition of naming boats after the women in the family. Your wife, your mother or your daughter. Danièle was the name of Charles' wife."

"And?"

"People wondered why it had a foreign name on the bow. The boat was called *Isabelle*."

I left the place where I was found, drove out of Le Crotoy past faded plastic bags flapping in the ditches. Followed the coast up to the Strait of Dover. I had travelled here with one question, and was coming away with two answers.

I pictured Einar at the time of liberation. The Daireaux family had been executed, Isabelle was missing. He must have known that his search for her would take some time. And that Winterfinch would soon be sending people to collect the walnut.

How would he have hidden a large consignment of precious timber, one that he believed belonged to Isabelle?

He would have built a boat out of it.

The last trial of this grief-stricken cabinetmaker, who had also learned to repair boats for the Shetland Bus, was to return to Authuille. Fell the trees with a loyal friend, drag the trunks out of the shelled woods along the safe path and transport everything to the coast.

He and Charles Bonsergent, men who ten years earlier had stood side by side at a carpenter's bench creating *art-deco* furniture, now positioned the walnut on the circular saw and cut the best materials they had ever seen. Made them in the shape of gunstocks, but a little larger, so that they could be fitted next to each other in a curved frame.

The wood would dry slowly in the damp coastal climate and would not crack. Einar named the boat *Isabelle*, so that she might find it if he died and perhaps understand the message that it carried. They set out and cast their nets, pulled in fish, earned money for a larger boat. Flipped *Isabelle* upside down on a sand bank in the unpopulated wastelands of the bay of the Somme. A wooden boat would need many days for the planks to swell and become watertight again, so there was little danger that it would be stolen.

If I had not wandered off in the woods, Winterfinch would have been reunited with the precious walnut stocks. Instead his life was set on a different course, and he dedicated what was left of it to Gwen. Out on Haaf Gruney, Einar began building coffins. The neighbours, such as there were, grumbled about this eerie sight in the twilight and told him to get another boat. A good reason to bring *Isabelle* home. The painted letters of her name had faded, and he rechristened it *Patna*.

I pictured it, two men in their late fifties crossing the Channel sometime in the early 1970s. Probably a fishing boat with a large rowing boat in tow, like a lifeboat. A perfectly ordinary sight. A manageable crossing. Four to five days, maybe, in good weather. Up the coast and across to Shetland. And in the end, some years after Winterfinch had died, Einar ended up underneath *Patna*.

An icy-cold wind blew in from the sea. On the Channel, the waves were white.

I thought about the weather back home in Norway. Snow would soon come. The sheep would be moving into the woods, they would not survive on the mountains in the cold and sleet. A disgrace for a farmer, to be the last person to bring them down.

No time to lose.

12

TWO DAYS LATER I WAS ON THE FERRY TO UNST. *GEIRA* rumbled onward, powerful and heavy. Visibility was bad, snow-flakes hovered in the air, the wind was cold and biting. I felt time pressing on me. If the weather was like this here, the sheep must be up to their bellies in snow at the mountain pasture.

Unst had become like a home to me. I stood at the prow of the boat, eating a chocolate bar from a vending machine whose whims I knew, remembered its particular clang when the coins dropped inside. The same men were at work behind me, wearing the same oilskins.

The Bristol was run down. The steering wheel rattled, the lights on the instrument panel had stopped working, a valve in the engine was making a loud ticking sound.

But as long as it got me to the boathouse, it could have its rest.

I ran down. The grass was wet, the stones dark with rain. A storm petrel flew above me. I grabbed the rope along the boathouse and swung towards the gate. The white cross shone against the grey.

The padlock was gone, a wooden stick was wedged in its place. There was no boat inside, just a hollow echo of sloshing water. I looked over to Haaf Gruney. It seemed infinitely far away, the island a deserted grey blot in the mist. As though nothing of what I had experienced had taken place. As though the photograph on Bestefar's roll of film had never existed.

Soon I found myself standing outside Quercus Hall. I had come across no cars, no people. Even the sheep had kept away.

The large house appeared older, and colder. The grass bent in the direction of the wind, a grey-brown leveret bounded towards its hiding place near the foundations.

Again I felt that a long time had passed. Ten, even twenty years.

I walked down towards the stone cottage. There, I finally saw something from the present: a bulging bag of rubbish on the doorstep. A plastic bag from Clive's Record Shop.

The door was open. Inside it was warm, but she was nowhere to be seen. A portable heater whirred in the living room. There was a kettle on the stove and next to it a cup. She had been sleeping on the sofa, curled in a blanket. I held it to my nose and recognised her smell. A perfumed scent I had last been aware of at Hôtel de la Basilique, which had gradually faded but now embraced me again.

I followed the path back down to the boathouse. The snow flurry picked up, flakes melted on my face. Then I spotted her. Out on the dark sea, oars raised, with *Patna* rocking in time to the waves.

She wore a black knitted cap. Her hair clung to her cheeks.

"Gwen!" I shouted.

She turned, did not seem surprised to see me. Only distant, dispassionate. From forty metres away I could tell that it was over.

Gwen dropped the oars and began to row. I would give her that, Gwendolyn Winterfinch, she had a natural affinity with boats, she was forged by the Shetland coast. This view of her could be four hundred years old, as old as the wood of the boat she sat in.

She was not rowing towards the boathouse, but towards a large, flat stone below me. Did she want me to climb on board? I heard the splashing of the oars as she pulled *Patna* level with the stone. I grabbed hold of the tow rope, put one foot in and kicked off with the other. I felt my body become one with the shifting sea.

"Didn't I tell you not to come after me?" she said quietly before looking up. I could not interpret her expression, it was as though she were struggling with what she should do next.

"Yes," I said.

"Then why are you here?"

"I wanted to see Unst one last time." I glanced at the tarred planks on the bottom of the boat, at the oars that had been worn smooth. "Why aren't you using *Zetland*?"

"Fancied a row. But maybe you were planning to row to Norway?"

I shrugged. There were only two reasons I had come back to Shetland so quickly: to see her, or to have *Patna* for myself. But the truth was, this time I wanted both.

"Where are you heading?" I said.

"How about Haaf Gruney? For old times' sake?"

She was no longer the furious, scorned girl who had left the hotel in France. Now she seemed in control of things, even a little manic, as though she were relieved at having come to some kind of resolution.

"Gwen," I said, "I think Einar—"

"I saw the lights of your car," she interrupted. "Never thought you were going to come back. I have no idea what you uncovered in France. But seeing as you're here, I'm going to give you something that belongs to you."

She released the left oar and reached into her jacket pocket, handed me something and started rowing again, faster now.

"Even the very first time we met I had this feeling," she said. "This strange feeling about you and me."

She had given me the wooden dog Einar had made in 1971. It was not how I imagined it; it was smaller and slimmer. But it was exactly like the dog in the photograph, and my hands remembered it too. The feeling of polished wood, the curve of the back, the joints in the legs.

"How in the world . . ." I began, and then I felt the past rushing towards me again. The dog had a push-up base with a spring, and when I pressed it, the head nodded, the legs collapsed, the tail wagged. My fingertips met something familiar carved into the wood. A squirrel hiding its nose in its tail.

"All my life I thought Grandfather had given it to me," Gwen shouted over the wind. She was short of breath, only spoke between strokes, and we were now far from the coast of Unst. "But he got this strange look when I held it, so I kept it in my room at Quercus Hall. I found it the day before yesterday amongst my toys. It *is* yours. That's why I couldn't stand being at the hotel. When I saw the photograph of you and the dog, the truth was impossible to avoid. Grandfather was there when you disappeared in the autumn of 1971, and I was there in the summer house."

We said nothing for a long time. All that could be heard was the groaning of the oars and the waves lapping at the boat.

"So you did have a summer house?" I said.

"I didn't remember it at all. After I left, I went to Amiens and had a good cry. Then I rang our business manager. He told me that the house was sold in 1972. He gave me the address and I got a taxi to drive me around. We found it in the end. Inhabited by an unemployed eccentric, it looked run-down and sad. When I stood outside it, I thought I recognised the smell of the earth in the garden."

Everything was falling into place now. But a deep regret about how I had treated her seeped into me.

"I have a hazy image of sitting on the floor playing with someone. You must have had the toy dog with you. I suppose Einar came at some point, and I must have taken your toy and hid it. Never had to share with anyone before. So it came with me to Shetland."

I pushed the middle of the base, so that all four legs collapsed and it lay down. By releasing the front it raised its head, like an excited dog when it realises that its owner is taking it for a walk.

"You need not be troubled by your family history," I said. "Your grandfather saved my life."

We were approaching Haaf Gruney. The breakers pounded the reef, and soon I would be able to see the buildings. I scanned *Patna*. The joints, the bottom planks.

I told her that I had been in the woods. That the bruises must have come from Duncan picking me up with his one arm. She listened, but her gaze wandered, it was as though nothing mattered any longer.

"Why didn't Einar go to the police?" she said distantly.

"I suppose the thought had not even occurred to him. What good would come of it? His daughter was dead. He knew he would have to give me up, and that his brother would hate him for eternity. He just wanted to spend as much time with me as possible."

"Do you think they forgave each other? Einar and my grandfather?"

"Who knows?" I said and balanced the dog on my knee. "Should I row a little?"

"No, no."

The precipitation alternated between snow and rain. The sea sloshed at the keel. Her hands were soaked, her skin puckered around the nails.

"Why are we rowing to Haaf Gruney anyway," I said eventually.

"Let's spend a few days there," she said. "You and I."

"I can't, Gwen. I have to go home and bring the sheep down from the mountains."

She looked over her shoulder at the island and adjusted her course.

"Then why are you here?" she said.

We said our goodbyes long ago, I thought. But how I answer her now will determine how we remember each other.

"Because this boat is made of walnut," I said.

She reacted strangely, seemed . . . disappointed.

"It must be," I said and told her that its name had been *Isabelle*. "That's why it's so difficult to manoeuvre. I assume the entire frame is made of walnut."

Gwen stopped rowing. She shifted her grip on the oars so one hand was free, pulled off her cap and ran her fingers through her hair.

"Not just the frame," she said. "The entire boat is walnut. Including the thwart you're sitting on. It carries enough wood for two boats, but you can't see that when it's in the water. The whole thing is a brilliant optical illusion. The keel is extremely deep, presumably made of thick walnut. The lower-grade wood is on the outside, and even that is remarkably beautiful. Three inches thick. The standard thickness for shotgun stocks."

We approached the southern side of Haaf Gruney. The stone buildings crept into view along with the sloping beach where driftwood landed.

"When did you work it out?" I said.

"When I came back from France. I was sitting in the stone cottage thinking about everything that had happened. I started thinking about *Patna*, that Einar had got it about the same time he started building coffins. A herring boat has to be sturdy. But *this* sturdy? I went down to the boathouse and scraped off a bit of tar."

"And are you sure?"

"Turn around and look under the thwart."

I got down on one knee. She had scraped and polished a patch of the blackened wood, and even in the half-light I could see it was just as beautiful as on the Dickson shotgun. It was as if she had cleaned a grimy window and revealed the view of a fairy-tale castle.

Gwen had been fiddling with something on the bottom of the boat, but now took up the oars again and held the blades high in the air, the water dripping off them. We rocked quietly past the sloping beach on Haaf Gruney. I sat down and tried to think ahead, to some kind of future.

"Let's share it," I said. "The trees grew on my grandmother's farm. The wood belongs to my family. But the pattern belongs to the soldiers."

She shook her head. The boat began to drift further from the island.

"That will only make the chasm between us more evident," she said. "And deep down you know it. If this were a normal boat, we would have driven to Lerwick, taken the ferry to Norway and brought down the sheep. I'm in no hurry. I'll stay for the winter if you want. But the money will always come between us."

"Then have it all," I said.

"That's easy to say now, but this old tub is worth a fortune. You could buy your farm twice over with the money. I could get Quercus Hall into shape, if not finish it. Money changes people, Edward. You tell me I can have it but you're lying to yourself. Nobody would just give it away like that. Not even you. You believe it belongs to you, to your family. You would keep an eye on my spending. Think I'm frittering away your inheritance. Hate every single handbag I bought. Ask yourself what you would have done if *Patna* had been in the boathouse when you arrived?"

"I don't know. Broken it to pieces, probably. And then . . . sat down and thought about it."

"Aha! It's what you'd do after you'd thought about it, that's where it gets interesting."

I remembered what Mamma had written in the visitors' book at Thiepval. "I think my mother would have liked me to sell it to your grandfather. With no fuss."

"*She* would, yes. But she was not the lover of a Winterfinch. You would soon have your doubts. At the first hint of an argument you would think you had failed Einar, failed the entire Daireaux family. That Duncan Winterfinch never kept his part of the bargain. At some point in your life you would need more money. You would curse yourself for having given it away, and you would curse me for having accepted it. But do you know what would hurt me the most?"

"Tell me. It sounds like there's a lot to choose from."

"Your mistrust. You'd always be thinking that I got involved with you in order to find the walnut. No, you never believed me, but do you believe me now? Can you feel your feet getting wet?"

I looked down. Water was seeping through the bottom boards. She pulled hard on the oars and the sudden movement of the boat made it splash against my shoes.

"What have you done?" I shouted in alarm.

"I don't want us to be forced to make a choice. I took out the drain plugs while you were looking under the thwart." She held up two thick brass screws.

"Have you gone mad?" I said, reaching for them. "Put them back!"

She dodged me, making the boat rock, and I had to hold on to the gunwale to keep upright.

Gwen tossed the bolts into the sea. The brass shimmered for a second before sinking into the darkness. I reached for the bailer, but it was gone.

"*Zetland*'s over there," she said, nodding towards the boat-house on Haaf Gruney. "I towed *Patna* over here, moored *Zetland* and rowed back to Unst. So we can get home after this."

The boat was already sitting lower in the water. Haaf Gruney was not far, but we were moving more sluggishly and drifting towards the black reef. Gwen seemed to have given up rowing altogether.

"Dear Lord, Gwen! You don't throw something like this away."

"Just as I said, now you're showing your true colours. Your little bit of greed."

"*No!*" I lunged for the oars. "We're not going to lose it. Is money all you see? There are four hundred years of history inside this wood, and a century of wars."

Then Gwen leaped up, twisted the oars free of the rowlocks and threw them into the sea.

"That's not how it's going to be. This boat is going to drift out to sea, and the two of us are going to swim to shore and start over. We'll dry off, make love and make this thing last as long as we can. The odds may be stacked against us, but I want it to work out!"

I scrambled for the oars, managed to pull one in and paddled towards the other. But it was drifting away, and the water level in the boat was rising.

"Let it go, Edward. *Patna* is heading for the bottom. A watery grave for a sad tale."

And then Shetland did what Shetland does best. The weather turned, a wind set in. The snowflakes transformed into a hard driving rain.

"Sit down!" I shouted. "Then I can row us in."

The water was now up to our ankles, our trouser legs had turned black.

"I want you," she said. "When I'm with you, my heart wants to beat its way out of my chest."

She stood up and quickly climbed onto the beam, and the sea poured in like the surface water of a dam. I threw myself to the other side to balance the keel, and she lost her balance and fell into the sea. The waves swallowed her, and the boat rocked heavily into place again.

I felt the other oar thump against the side of the boat. I pulled it in, slotted both oars into the rowlocks and tried to row after her. But the sound of the oars between the pins had changed from a dry creaking to a heavy, obstinate scraping. The waves were now cascading in.

I pulled the oar loose and slammed it against an oar pin, over and over until the pin cracked. I broke it off and stuffed it into the drain hole, hammering it in place with the oar, and did the same with the other. But it was too late. The sea belched in with each movement I made – *Patna*, the sea and I were one.

I kicked off the hull and swam after her. The water was ice-cold and my jacket slowed me down. I saw the wooden dog drifting past, only its head above the water. Gwen had gone quite some way, there were only a few metres between her and the shore. But the waves grew higher and more restless, it was difficult to swim in.

I glanced back. *Patna* was still suspended at the water's surface, like a full tub. But now Gwen was heading towards the nearest land, towards the reef where the sea was churned into foam.

"Not there!" I screamed, swallowing salt water. "You have to get to the beach!"

A wave had carried her there, and when she was a few metres away, another came in and drove her hard against the black skerries. There was nothing for her to grab hold of. The crashing water muffled her screaming and immediately I saw blood, washed away by the next wave.

13

DRY SNOW DRIFTED THROUGH THE CRACKS AND FORMED small, white fans on the floor. On the windowsill there were empty flowerpots with traces of soil at the bottom. Alma would have put them there, and Bestefar must have removed the plants when they withered after her passing. Nobody had made any effort to brighten the place up since then. The cabin at the mountain pasture refused to get warm. I re-filled the cast-iron stove and shivered. A fly had come to life and crawled towards the stove, and now sat there bewildered.

I heard the bleating of the sheep outside. Twelve so far, all with clumps of ice stuck to their shaggy wool. They were barely able to drag themselves through the snow. They didn't like harsh, windy weather, and most were terrified and had migrated to the woods. They ran for dear life when I approached, even though they were starving.

I went back outside, and a few hours later I had found only one more ewe. They had begun to wander off on their own, and were so afraid of people that it was a struggle to get them to follow me home.

I stood near a patch of felled forest. I could see Saksum in the valley far below. The last time I had drawn attention to myself in the village was when I had arrived in the Bristol with a foreigner. We had bought Worcestershire sauce as a kind of protest. But it was a carnival; the car too odd, the jacket too much like a disguise, the explanation too longwinded.

Now I wanted to be normal on the outside, so that I could get on with my real life on the inside. I wanted to go to the café at the railway station without being stared at. I no longer wanted to be the person who chases after people when they paint swastikas on cars. I wanted to go trapshooting without anyone wondering what I might be doing with a shotgun. All perfectly normal activities, which each and every person can have. I wanted to go down to the village and feel that I belonged, and at the same time be myself.

So what should I do? Just *do*, I guess. Drive to the shooting range with my Dickson Round Action. Shoot a couple of series, pack away the gun and leave. Let people talk. Come back the next time. Just that. Business as usual. Let them see me.

That was it, really.

I rounded up a few more ewes. Their wool was full of twigs, a nuisance come shearing time. They sped up when they heard the bleating of the rest of the flock. Near the fence, I couldn't work out the numbers. There were more than twenty now. Someone had chased them in, but there was not a soul to be seen. Footprints by the gate, but I had not heard a car.

When I went back the next morning the same thing happened. Someone had been by the pen chasing in more sheep than I had, while I was down in the woods.

The real cold arrived in the weeks that followed. Snow settled on the fields and the woods, sparkling, fluffy snow that covered the branches of the trees. I worked from seven to seven, to block out all thought of what had happened in Shetland.

In the loft I found the old chart of freshwater fish. It had been included as a supplement to the summer edition of a weekly magazine. I had hoped that seeing it again would bring back more of what Pappa had said to me, but his voice was no more than an echo.

Maybe that is how it is, I thought. We are not meant to remember everything.

I cleared a path to the workshop, fired up the Jøtul stove and made a wooden frame for the chart. I went back up to the drawing room on the second floor, cold through the winter, and dusty and barren in the summer. Each generation seemed to haunt this place; on the walls were small photographs of gruff-looking men and subdued women with pinned-back hair, it was as though they lived up here with no visitors, abandoned and unwanted by those who came after.

This is how empty a farm can get, I told myself. When the memory of Bestefar's living room is the only thing that can warm me. Him standing in front of all those books, the glow of the Grundig stereo equipment, the amplifier's thin needles flicking in time with his hands as he conducted absent-mindedly, and through it all, the smell of roast meat served at five o'clock on the dot, two plates, and the smoke of his cigarillos.

Up here there was nothing, just the cold floor of a large, uncarpeted room, a dinner table for twelve that had never been used, leather-bound books with gothic lettering.

I hung the chart in the hallway of the cottage. Cooked a trout from the summer fishing trip and shared it with Grubbe.

The wheels spun as I drove down the dead-end road. His driveway was so poorly cleared that I had to stop outside the gate. The Rover was in the carport with its summer tyres still on. The longest expedition for him in the winter was probably down his path, from the front door to the postbox.

He was not dressed, even though it was twelve o'clock, and he seemed thinner and more stooped than I remembered him.

"Have you eaten since the summer?" I said.

He was about to reply, but broke into a coughing fit and waved me inside. "Nothing has flavour anymore," he said and shuffled in after me. His slippers smacked against the worn parquet in the living room.

"Coffee?" I said. "I'll make a pot."

"You have the blessing of the Church. And would you mind

fetching the newspaper? Even though it only gets worse. I'll have to switch to *Dagningen* soon."

Apart from teabags and a packet of biscuits his kitchen cupboards were empty. There were crumbs and a dirty cup on the kitchen table. I heard him coughing in the other room.

I drove to the general store and shopped as if I was going to spend the winter at the mountain pasture. I picked up copies of *Aftenposten* and *Vårt Land* and filled two boxes with food.

When I came back into his hallway I heard Radio Oppland on full blast. I put the food away and made coffee in a large pot I found in the cupboard. The priest was of the generation that used one large enough for an entire evening of T.V. I followed the direction of the sound to find him lying in his office, on a couch beneath a wall-mounted bookshelf.

"Do I smell coffee?" he said, getting up. "Or has He carried me home?"

"Thankfully it's only coffee, the red Co-op brand," I said.

He switched off the radio.

"That will do. Have you made plenty?"

"One and a half litres."

"Good," he said. "Now tell me everything. Every last drop."

When I stopped talking he sat for a long time with his eyes closed. His body rocked slowly back and forth.

"Are you asleep?" I said.

He opened his eyes. "Not at all. Rarely have I ever been more awake."

The old priest thrust his chest out to stretch his back. "As the priest in Saksum I have heard and seen a great deal which has made me doubt whether the Good Lord's plan is actually beneficial to mankind. But that does not seem to be the case today."

Outside, the winter darkness had settled in.

"Does it torment you?" he said. "That you ran into the woods."

"Yes," I said.

There was silence.

"You must let it go, Edvard. I have no scripture to offer. But we are innocent when we dream and we are innocent when we are small."

"I know that," I said. "But I can't seem to wriggle my way out of it."

"You have served a sentence many times over, one which nobody has judged you for. Think of your grandfather. I see him in a clearer light now, sitting alone at the organ concert. Maybe he knew everything that had happened and was afraid that you, when you missed them the most, would blame yourself for the death of your parents."

We were briefly silent again. His chair creaked as he twisted in it. "Why do we actually call it a television pot?" he said. "There's no such thing as a radio pot."

We shared the dregs of the coffee.

"These two women interest me," the old priest said. "Which one made your heart beat stronger?"

"I wonder that myself. It was like it was beating for Hanne when it pumped in, and for Gwen when it pumped out."

He got up stiffly from the chair. "I had a girlfriend once, when I was at the seminary. But it came to nought. I hesitated, I wasn't quick enough when it mattered."

His eyes scanned the bookshelves, the piles of old sermons. "Take a look inside the green box by the desk."

I crouched down and flipped through the yellowed proofs of the Saksum parish newsletter.

"I made them with Letraset and a typewriter. But dig down a little further and you'll see something which is much closer to God. And then you must promise me that you will not shut yourself away."

There was a well-thumbed stack of *Playboy* Special Editions.

"As a bachelor, I have had to make myself familiar with such 'celibacy publications'," the old priest said. "The regular editions are not that special, I don't care for their hedonism. But every May there's a special edition. *Girls of Summer*. I've been a subscriber for years. As you can see, it is all girls and no words. Nothing explicit. Just creation's most magnificent *opus*. Woman as she was given to us by God. Unfortunately no paper edition of Woman will ever live up to the real thing. For the Holy Bible, however, it's good enough."

It was early February when I walked up to the flame-birch woods. Alone with an axe, no tobacco, no packed lunch. I chose the first one I came to, held the axe sideways and started to chop. A sprinkling of snow fell from the branches. I looked up at the sky, kept my eyes open and let the snowflakes melt on my cheeks.

I stepped around the stem and made an even notch on the other side so that the birch balanced on a thin wedge, like a half-empty hourglass. I placed my hand on the tree and gently pushed it, as though I was teaching a child to dive, my hand on its back to show encouragement. With a drawn-out creak the flame-birch toppled, the deep drift gently welcomed it, and a muffled multitude of branches snapped beneath the snow.

I limbed the tree, measured out lengths, chopped it up with a handsaw and fetched the old Deutz. By the time blue dusk arrived, I was towing eight logs. I drove out of the woods in a low gear and saw the farm open up before me. It was still encircled by a hazy winter light, the snow packed between the notched trunks of the log house.

For three days I worked like this, with only an axe and a handsaw so I would not need ear protection. A cold, hard slog which held thoughts at bay. Some of the iron bands snapped as the birches fell, others I broke off with a chisel when I brought the trunks to the farm, and I gathered them in a jagged pile of scrap iron.

On the fourth day I was as good as finished with the limbing and driving when a heavy snow began to fall. A swirling wind settled in, and through the flurries I glimpsed a pale-red flash between the tree trunks. She stepped onto the path I had made, her blonde hair turning chestnut-brown again.

"The potatoes up early and the sheep down late. You are foundering as a farmer, Edvard Hirifjell." She stopped five metres away. "But perhaps things haven't been that simple for you."

"Thanks for helping with the sheep," I said.

"Yes, that's me, Solvoll the veterinarian."

Her freshly oiled mountain boots turned the snow to slush. She climbed onto a birch trunk. The leather boots offered a sharp contrast to the white birchbark as she rocked and balanced, as though she were guiding it downstream.

"Are you really going to chop down all the trees?"

"That's their destiny. And mine," I said.

"I'm not sure it's your destiny," she said, hopping down. "Look at which one you've left standing."

Hanne went to the large birch we had lain under the previous summer. The bark was furrowed and rough, the iron bands embedded. She ran her finger over the rusted metal.

"You're standing here in the same work clothes I've seen you wear so many times," she said. "Yet you've changed completely. I don't know whether I like you or not. And I really don't know if I want to find out."

I let the axe fall into the snow.

"I'm never going to be some poor girl who waits faithfully in the village. I've accepted a two-year internship in the north. So you're not getting something for nothing, Edvard. But you don't owe me either."

Two years. In some respects it was shorter than two months.

"Aren't you going to say something?" she said.

"I'm trying to think of something nice to say to you," I said. "But there are too many nice things to choose from."

I picked up the axe by the end of its handle and let the head swing back and forth, drawing two connected circles in the snow.

"What's that?" she said.

I asked myself the same question. Was it the sign for infinity? Or just a figure of eight?

She twisted a strand of hair around her finger. "What happened to that other girl?" she said.

I lifted the axe a little, still swinging it in the same pattern but above the snow. She could not have known what had happened. But I could tell her.

"If you want to talk about her," I said, "don't call her 'that other girl'. Call her Gwendolyn Winterfinch."

14

THAT DAY OUT ON HAAF GRUNEY I HAD SWUM AFTER her, grabbed hold of her arm and tried to find purchase on the rocks. My entire body ached with a dull pain, my skin was stinging, and one hand was bleeding from having gripped a sharp stone to try to haul us ashore.

It was impossible. I lost my grip and she was pulled out to sea again, unconscious. She was tossed back and forth in the waves like a rag doll, her hair spread in a fan, and she drifted off before the next swell lashed us once more against the rocks. She was thrust into my arms, I gripped her and swam on my back towards the sloping beach, trying to keep her head above water.

A wave took us in, and I crawled with her onto the round stones. My sweater hung heavy, water ran from my hair. When we were further up I tried to revive her. Heart massage, just as they taught us in national service. Mouth to mouth. Firm pressure on her chest. She gagged, tried to raise her arm, but it fell limply, like a severed branch. I remember noticing with some surprise that Duncan Winterfinch's watch was still working, attached to a lifeless arm.

I manoeuvred *Zetland* out of the boathouse, carried her on board and went full throttle to the ferry dock at Yell. All I could think of was her funeral, that I would have to persuade her family to lay her to rest in the black coffin in the outbuilding on Haaf Gruney. The next thing I remembered was a large hand on my shoulder and two doctors tending to my injuries.

Three days later she was conscious again. But by then she was already lost to me. It was as though she had been lifted out of our dreamlike world into reality, onto a sick bed, into the doctors' embrace. She belonged to them. To them and to the family that encircled her bed, a wall of expensive clothes and anxious looks. They did not want to talk to me, only wanted to hear from the sheriff what I had said. They would keep it within the family, the story of how the thoroughbred broke free and spent the summer with a mongrel. I wanted to speak, and I wanted to hear them talk about her, about this person I only knew from my own impressions of her. But the doctors simply shook their heads and asked me to leave.

I took *Zetland* back to Haaf Gruney and waited there for someone to claim the boat back from me, threaten me with lawyers, hold me responsible. But nobody came and the house grew dark.

The next morning I sat on a rock looking at *Patna*. The same current that carried driftwood to Haaf Gruney had brought it into the shallows, where the tide abandoned it like a stranded whale. The keel was scratched, it looked like a cadaver and smelled of tar and salt water.

I could bear it no longer. I went back to Lerwick, to her hospital bed, and found her alone.

"Is it you?" she said, adjusting a plastic tube stuck with a plaster to the back of her hand.

"Yes. It's me."

She looked towards the window. They had cut her hair short and shaved it around the wounds to stitch up her scalp. A dressing ran from her ear to the base of her neck. Two large plasters.

"I will make things right again," I said. "I'll follow you wherever you want to go. Just let me make things right again."

Her eyes were glazed, her lips cracked. There was a pervasive smell of antiseptic.

"Do you need anything," I said. "Water? Something to eat?"

"Why don't you ask if I need *you*?" she said drowsily.

"I'm asking you that now."

"I need you. But I can never have you." She coughed, then paused to collect her strength. "When I lay here alone, I hoped you would come and say these very words. But no. You're a wounded animal. When you're back to full health, you'll run off and I'll never be able to keep up with you."

"Don't say that. Please forgive me for doubting you."

"Dear Edward, I meant everything I said in the boat. But in the end, it was the adventure that kept us together. Let me be spared that vile day when I realise our relationship won't work out."

She reached out a hand and placed it on my forearm. My skin tingled.

"Do you know what I've been most afraid of all my life, Gwen?" She shook her head faintly.

"That I would turn out cold. Emotionless. That I would never feel anything, or grieve for anyone. I am no longer afraid of that."

"Then we feel the same," she said and closed her eyes.

I took her hand and held it until a nurse asked me to leave.

Back on Haaf Gruney I slept until I was woken by the cold. It was a bright, clear morning, and through the window I saw *Patna* capsized in the shallows.

What would I have done if Gwen had drowned? I might have doused *Patna* with petrol and set it alight. They had been real, my feelings for her. Real and true. Because I *was* able to imagine the two of us, later in life. Hopeless, but true nonetheless.

I cried so hard that my stomach hurt, my tears falling onto the dust. I was angry with her for not wanting to try, angry at myself for not throwing myself in after her straight away and letting *Patna* fill with water. I imagined her with other men and wondered how she would tell them about me.

We had been hard on each other. Not hard in the sense that we lacked feelings, but what each of us carried inside was hard. I remembered a paragraph from one of Einar's letters in which he described how they amused themselves during their lunch breaks at Ruhlmann's workshop. They would compete to cut two blocks of wood so straight that a drop of water would hold them together. Maybe that is how it was between Gwen and I. Close to each other, locked in place by surface tension.

I walked down to *Patna*. The water was still dripping out of it. The wind carried a bitter, raw cold. Out at sea, a shower was on its way.

The dead gathered around me and said:

The time has come. What are you going to do now?

It was Bestefar I listened to. The man who would come home to listen to Bach after spending an entire day digging up rocks.

The wood came from the earth. It was like failing to harvest corn. Or neglecting to bring in the sheep. Or not digging up the potatoes.

I thought back on the two of us. The harvest was our time. Between withered potato vines that had absorbed light and nutrition for the crops in the darkness beneath the earth. The droning of his two-wheeled tractor. My hands in the black earth that got under my nails, my knees damp as I picked out the potatoes and placed them in the wooden crates, while Bestefar walked ahead, sharing the same silent satisfaction. Our sheltered world, harvesting potato after potato and setting aside the best for Christmas.

The memory ended, and I began.

I fetched the saw and crowbar from the workshop and tried to prise off some of the planking. But it was hard as iron and refused to give. My ribs stung where I had been injured, and for an entire day I failed to reverse any of Einar's meticulous boatbuilding. Each length of wood was nailed and coupled to withstand collisions and the unrelenting sea. *Patna* had a locked and sealed construction, a pyramid with no entrances, and if I was too rough with the crowbar, the wood just chipped.

The next morning I felt Einar's cabinetmaking begin to talk to me. The hull of *Patna* creaked and moaned, until at last I loosened a couple of boards and knocked apart some of the planking. I was less heavy-handed and haphazard than I had been the previous day, trying to think as he would have thought.

From then on Einar was with me, his hands became my hands, his plan became my plan, his thoughts entered my brain, and soon I knew where to place the crowbar to unlock the joints and pull away plank after plank. I soon grasped that the frame, the boat's ribs, consisted of several joined half-metre lengths of wood. They stretched symmetrically into the centre, black and rough.

For five minutes I just stood and stared. And then I broke loose two gunstock blanks, perfectly cut and covered with tar.

I worked for the next three days, one in rain, the following in the sun, the third in snow. In the last of the afternoon light, with the gulls shrieking from the nesting cliff on Fetlar, I carefully tore loose the last timbers. *Patna* was no more.

All that remained was a pile of stock pieces. Twenty high, sixteen across. Using a hand plane I scraped off the outer layer of tar, fetched some sandpaper and rubbed down to the wood.

A light flurry came in from the east. I lifted my face to the sky and allowed the flakes to cool my forehead. The snow settled on the walnut and melted, and within seconds the wood had absorbed the moisture and brought out the colours and pattern.

I transported the walnut across the sound in *Zetland* and let myself into Quercus Hall. I walked through the long corridors smelling of mildew and old furniture polish, and carried the wood up to Duncan Winterfinch's office. I stood there for a while, breathing in the stale smell of Balkan Sobranie. I studied the wide photograph of the Black Watch and looked out over the crowns of the trees in the arboretum, now leafless and naked.

I turned to the walnut one last time. I could see centuries of the Daireaux woods, the meandering black and orange forms. Only then, in the half dark, did I notice that the wood was faintly luminous, like the hand of a watch that releases the daylight it slowly measures throughout the day.

It was as if the pattern displayed everything the trees had

witnessed for four hundred years. But it also offered a glimpse of something endlessly deep within, shades of colour so unpredictable that they were not of this world.

It was the light that shone from the kingdom of the dead, which they had all stepped inside.

The soldiers of the Black Watch, Isabelle, Einar, Duncan Winterfinch, Bestefar, Mamma and Pappa.

V

Isabelle

THE YEARS PASSED. YEARS OF BLIGHT, YEARS OF RECORD harvests, years of drought. The new tractor became the old tractor, the old tractor was relegated to the barn.

We lived with the earth and the seasons, with plants that blossomed and withered. But within me a slower cycle rotated, a maturing that needed many years to come into bloom.

One day, at the end of a warm May, I realised that the time was ripe. I was up by the county road surveying Hirifjell. The post had not arrived, I had listened for cars, but everything was quiet, just the faint roar of Laugen in the valley and the wind whistling through the trees.

It was time. I took the shortcut past the stinging nettles and into the workshop, where I pulled out a box wrapped in sailcloth.

She was sitting at the kitchen table, text messages flying back and forth. "I have something for you," I said. "But you'll need better shoes."

We walked through the gate, through the potato fields, past ground newly planted with summer cabbage, up to the flame-birch woods. I still called them that, even though only one of the trees was birch. Around it grew walnut trees, sixteen in all, from the nuts I had collected in the Daireaux woods. They had germinated in pots and should have withered the first winter after I planted them out, but somehow they survived. They swayed in the wind but held their own up here, on the sunny side on the inside of the far side.

*

My last act before leaving Haaf Gruney had been to drag Isabelle's coffin out from the peat, wipe it clean and polish it. In Lerwick I had asked at the funeral parlour whether they could look after it until it was needed.

When Agnes Brown died some years later, she was buried in Norwick, next to Einar, the psalms struggling to be heard above the gusts of wind. We were a small gathering. Inside the church, the coffin was surrounded by orange tulips and white lilies identical to those Einar had carved. Only I knew that Agnes would be buried in a coffin intended for my grandmother, and that it held a woman who had loved Einar, at last. He had been a living ghost, a man who restlessly hunted for a level of perfection only achievable at the carpenter's bench. As we lowered the coffin, I hoped that he, in another time, in another place, could reciprocate what Agnes had wanted to offer him. Just as in another time, in another place, Gwendolyn Winterfinch and I could have been together.

As we stood by the grave I saw her figure coming down the path from which you could see the lighthouse on Muckle Flugga, and I noticed she was carrying flowers. She sat on the hillside and followed the ceremony from a distance.

I hoped that the last verses of *"Kjærlighet fra Gud"* would be carried to her on the sea breeze, and that the words would apply as much to her as they did to Agnes, and that they should apply to everyone who came after us.

And the words would prove to do just that. Because now, standing in the flame-birch woods, I unwrapped the sailcloth from the chest and handed it to my daughter. She said nothing for a long time. It is one of her many qualities, to allow her reactions to ripen fully before revealing them to the rest of the world. As she held the chest up to the light, sunbeams penetrated the amber-yellow wood and revealed the infinite threads in the pattern.

"It's flame-birch, from these very woods," I said.

"Really! Is it a jewellery chest?"

"The chest is not the present. You have to look inside."

The sound of the lid opening drifted through the tree trunks. Inside there was a radiance from a wood deeper and wilder and older than the flame-birch. It contained a package in grey silk paper.

She knelt to rest the chest on the ground and opened the package. When she stood up again, her movements unfolded a shimmer of navy blue.

"A dress?" she said.

"Yes," I said and took the chest, which now felt strangely empty.

"So pretty," she said, holding the dress in front of her. "Where did you get it?"

"It has always been in the family."

She held it up to the sunlight and walked amongst the trunks to the old birch, wide enough to get changed behind.

When she reappeared, she had filled the dress with life.

The leaves of the walnut trees rustled in the breeze. But above their crowns stood the large birch tree, tall and unshakeable, with branches so thick that they did not sway. And there, through the foliage, the sun cast its perpetually shifting patterns of light and shadow upon us.

IN THE LATE 1990S, THE BUILDINGS ON HAAF GRUNEY were destroyed by a furious gale. Today the island is a nature reserve and home to a colony of storm petrels, who, for reasons unknown, orbit the churchyard at Norwick before they set out to sea.

LARS MYTTING, a novelist and journalist, was born in Fåvang, Norway, in 1968. He is the author of *Norwegian Wood: Chopping, Stacking and Drying Wood the Scandinavian Way*, which has become an international bestseller and was the *Bookseller* Industry Awards Non-Fiction Book of the Year in 2016. His novel *Svøm med dem som drunker*, now published in English as *The Sixteen Trees of the Somme*, was awarded the Norwegian National Booksellers' Award and has been bought for film.

PAUL RUSSELL GARRETT is a translator from Danish and Norwegian of novels, plays and books for children. He is also Programme Director for a new theatre translation initiative, [Foreign Affairs] Translates!